MODELING AND ANALYSIS OF DIFFUSIVE AND ADVECTIVE PROCESSES IN GEOSCIENCES

MODELING AND ANALYSIS OF DIFFUSIVE AND ADVECTIVE PROCESSES IN GEOSCIENCES

Edited by W. E. Fitzgibbon
University of Houston

Mary Fanett Wheeler
Rice University

siam.

Philadelphia

Society for Industrial and Applied Mathematics

MODELING AND ANALYSIS OF DIFFUSIVE AND ADVECTIVE PROCESSES IN GEOSCIENCES

Library of Congress Cataloging-in-Publication Data

Modeling and analysis of diffusive and advective processes in
    geosciences / edited by W. E. Fitzgibbon, Mary Fanett Wheeler.
        p.   cm.
      Includes bibliographical references (p.   ).
      ISBN 0-89871-299-8
      1. Fluid mechanics.   2. Porous materials.   3. Geophysics.
  4. Earth sciences.   I. Fitzgibbon, W. E. (William Edward), 1945–
      II. Wheeler, Mary F. (Mary Fanett)   III. Society for Industrial
  and Applied Mathematics.
  QC155.M63   1992
  550'.1'5—dc20   92-16400

PREFACE

In September of 1989, the Society for Industrial and Applied Mathematics, with the cooperation of the Department of Mathematics and the Energy Laboratory of the University of Houston, hosted a conference on mathematical and computational issues in geophysical fluid and solid mechanics. This was the third in an ongoing sequence of SIAM conferences pertaining to the geosciences.

The purpose of this conference was to provide a forum where mathematicians, geophysicists, geologists, hydrologists, and petroleum engineers could meet, discuss, and collaborate on problems of mutual interest. The central topics were systems of conservation laws, reactive flows, fluid and solid mechanics, partial differential equations, wave propagation, materials response, and geochemistry. Areas of application include flow in porous reservoirs and acquifers, basin modeling, seismic modeling and inversion contaminant transport, and remote sensing.

The program of the conference consisted of invited plenary lectures, minisymposia or invited special sessions, research workshop contributed papers, poster sessions, and informal seminars and discussions.

These volumes are not intended to be a proceedings of the conference per se. Participants in the conference were encouraged to submit manuscripts developed from the topics presented at the meeting. The submissions could either be expository papers or original research. The standard elongated abstract was not acceptable. All papers were refereed by outside reviewers.

The accepted papers fall roughly into three categories which we have somewhat arbitrarily labeled as Computational Methods in Geosciences, Modeling and Analysis of Diffusive and Advective Processes in Geosciences, and Wave Propagation and Inversion. These broad categories serve as the titles of this sequence of three volumes. We realize that our categorization is imperfect and hope that it does not serve to confuse or offend.

If dedications are appropriate, these volumes should be dedicated to Garrett T. Etgen, Chairman of the Mathematics Department at the University of Houston, who was tremendously supportive of the endeavor, worked endlessly, and derived no credit or visibility. In this dedication we show our deepest appreciation for his tireless efforts to further the development of mathematical sciences in the Houston area.

The errors that most assuredly occur in these volumes are consequences of the carelessness and incompetence of the editors and the editors herewith and henceforth apologize.

W. E. Fitzgibbon
Mary Fanett Wheeler

CONTENTS

CHAPTER 1

Numerical Simulation of Diphasic Flow in
Heterogeneous Porous Media by Homogenization*

B. Amaziane**
A. Bourgeat†
J. Koebbe‡

Abstract. A mathematically rigorous method of homogenization is presented and used to analyse the equivalent behavior of transient flow of two incompressible fluids through heterogeneous media. Asymptotic expansions and H-Convergence lead to the definition of a global or effective model of an equivalent homogeneous reservoir. Numerical computations to obtain the homogenized coefficients of the entire reservoir have been carried out via a finite element method. Numerical experiments involving the simulation of incompressible two-phase flow have been performed for each heterogeneous medium and for the homogenized medium as well as for other averaging methods. The results of the simulations are compared in terms of the transient saturation contours, production curves, and pressure distributions. Results obtained from the simulations with the homogenization method presented show good agreement with the heterogeneous simulations.

1. Introduction. We have rigorously derived "homogenized" or "global" reservoir equations from exact local incompressible, immiscible, two-phase flow equations in a periodic heterogeneous medium for a single rock-type model including capillary and gravity effects, [1]-[6]. The homogenization of a miscible model

*This research was supported in part by a U.S.-France Cooperative Science Program (NSF/CNRS).

**Laboratoire de Mathématiques Appliquées, URA-CNRS 1204, Université de Pau, Av. de l'Université, 64000 Pau, France.

†Equipe d'Analyse Numérique, URA-CNRS 740, Université de Saint-Etienne, 23 rue du Dr. Paul Michelon, 42023 Saint-Etienne Cédex 2, France.

‡Department of Mathematics and Statistics, Utah State University, Logan Utah, 84322-3900, USA.

without any dispersion term has been studied in [7],[8].

For the purposes here we assume that the local porosity $\Phi^\varepsilon(x)$ and the local absolute permeability tensor $K^\varepsilon(x)$ are rapidly periodic oscillating functions in the space variable. Then by homogenization methods: two-lenght scales asymptotic expansions [9],[10] and H-Convergence [11], we get global or effective permeabilities K^h and porosity Φ^h. In the multidimensional case, coefficients of the homogenized permeability tensor, K^h are obtained by solving linear partial differential equations on a basic cell with periodic boundary conditions.

In this technique, the assumption of periodicity looks rather restrictive. But we must admit it is by now the only technique which allow us to address rigorously, nonlinear phenomena such as the two-phase flow system. We must also recall that this assumption of periodicity could be relaxed, since a non-uniform periodicity (i.e. coefficients depending on both the period ε and the space variable) is sufficient to get analogous results and to account for a wide range of physical heterogeneities. We must also recognize that in case of linear convection/diffusion equations, local stochastic heterogeneities assumptions give a result akin to this one obtained from deterministic assumptions [12]. In that case, moreover, it could be proved that homogenization theory gives analogous results to these obtained by volume averaging [13]. In the same way, numerical results of multiscale technique with periodicity assumptions are in good accordance, with experimental data for various heterogeneous media [14].

2. The Two-Phase Immiscible Model. Standing governing equations modeling two immiscible fluid phases with no mass transfer in a porous reservoir $\Omega \subset \mathbb{R}^n$, $1 \leq n \leq 3$, over a time period $J =]0, T[$, [15]-[16], are made up of equations describing conservation of mass in each phase coupled via a capillary pressure law $P_{cap}(x, S)$, Figure 1, and relative permeablities curves $k_{ri}(S)$ in each phase $i = w, o$. S_i and P_i are the saturation and pressure in each phase, where the subscripts w and o refer to water and oil, respectively.

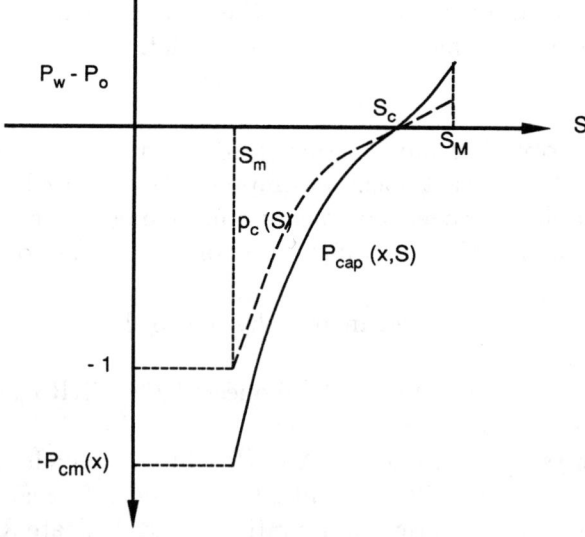

Figure 1. Capillary pressure curve at fixed x, $P_w - P_o = P_{cm}(x)p_c(S)$.

The equations of conservation of mass and continuity for each phase are given by:

$$\Phi(x)\frac{\partial S_i}{\partial t} + \nabla.\varphi_i = 0 \qquad x \in \Omega, t \in J \tag{2.1}$$

$$S_w + S_o = 1 \tag{2.2}$$

where φ_i is the velocity which is determined by the traditional extension of Darcy's law:

$$\varphi_i = -\frac{k_{ri}(S_i)}{\mu_i}K(x)\nabla(P_i + P_g) \qquad i = w, o \tag{2.3}$$

where μ_i is the viscosity of phase i, P_g the gravity potential, $K(x)$ and $\Phi(x)$ the absolute permeability tensor and porosity of the material.

The pressures in the two phases are related by the capillary pressure [17]:

$$P_{cap}(x, S) = P_w - P_o = P_{cm}(x)p_c(S) \tag{2.4}$$

where $P_{cm}(x) > 0$ is the maximum of the absolute value of the capillary pressure at the point x and $p_c(S)$ is a dimensionless function such that $p'_c(S) > 0$, Figure 1.

The system (2.1)-(2.4) may be rearranged in a form which very closely resembles to the system describing the displacement of two fully miscible fuids, [15]. This form is more suitable for mathematical purposes and will allow us to get results for both miscible and immiscible fluids as in [18],[19].Let:

$$\lambda(S) = \frac{k_{rw}(S)}{\mu_w} + \frac{k_{ro}(S)}{\mu_o} \qquad \text{the total mobility}$$

$$\lambda_i(S) = \frac{k_{ri}(S)}{\mu_i \lambda(S)} \qquad \text{the phase i relative mobility}$$

The system is now rearranged using the "global" pressure variable [18], [19] given by:

$$P = \frac{1}{2}(P_w + P_o) + \frac{1}{2}P_{cm}(x)\gamma(S) \tag{2.5}$$

where $\gamma(S) = \int_0^{p_c(S)}(\lambda_w - \lambda_o)(\tau)[p_c^{-1}(\tau)]d\tau$. Combining equations (2.1)-(2.4) we obtain:

$$q_t = -\lambda(S)K(x)\nabla P + \lambda(S)[\gamma_1(S)q_1 + \gamma_2(S)q_2] \tag{2.6}$$

$$\nabla.q_t = 0 \qquad \text{in} \quad \Omega \times J \tag{2.7}$$

$$\Phi(x)\frac{\partial S}{\partial t} + \nabla.(R + \lambda_w(S)q_t + b_1(S)q_1 + b_2(S)q_2) = 0 \tag{2.8}$$

where
$q_t = \varphi_w + \varphi_o$ is the total fluid velocity.
$R = -P_{cm}(x)K(x)a(S)\nabla S$ is a diffusion term due to the capillary pressure.
$q_1 = -K(x)\nabla P_{cm}(x)$ is a convective term due to the capillary pressure spatial variation.

$q_2 = -K(x)\nabla P_g(x)$ is a convective term due to the gravity. Also

$$a(S) = \lambda(S)\lambda_w(S)\lambda_o(S)p'_c(S)_{\geq 0}, \qquad a(0) = a(1) = 0$$

$$\gamma_1(S) = \int_{S_c}^{S} \lambda'_w(\tau)p_c(\tau)d\tau$$

$$\gamma_2(S) = \lambda(S)\frac{\rho_w\lambda_w(S) + \rho_o\lambda_o(S)}{\rho_m}$$

$$b_1(S) = \lambda(S)\lambda_w(S)\lambda_o(S)p_c(S)$$

$$b_2(S) = \lambda(S)\lambda_w(S)\lambda_o(S)\frac{\rho_w - \rho_o}{\rho_m}$$

with ρ_i, the phase density of the ith fluid and $\rho_m = (\rho_w + \rho_o)/2$ the mean density.

The equation (2.6) is frequently called the "global" pressure equation. It is a family of elliptic equations (one for every $t \in J$) with saturation dependent coefficients. The equation (2.8) may be considered a saturation (or concentration) equation, it is usually a convection-dominated parabolic equation for saturation (or concentration) with coefficients depending on pressure through the total Darcy velocity. In the case of immiscible displacement, the diffusion term R in the saturation equation vanishes at some points ($S = 0$ or $S = 1$). The equation degenerates at these points.

3. General Presentation of Homogenization. Let u be a quantity describing a process with two different scales. This could happen for instance with any process (heat conduction or fluid flow, plate bending, etc...) taking place in a heterogeneous region Ω. In this case, there is a small scale which is the scale of the heterogeneities l and a large scale which is the scale of the region Ω denoted by L. The quantity u will then depend strongly on these two scales and more precisely on their ratio $\varepsilon = l/L$.

The purpose of the homogenization method is by means of mathematical analysis, find the homogeneous properties of the region Ω which give a "globally equivalent" process through a homogeneous medium. This "globally equivalent" medium will be called "homogenized" and the parameters characterizing this medium will also be called "homogenized" or "effective" coefficients. From a mathematical point of view, to homogenize is simply to find the equation satisfied by u_ε as ε tends to zero, and thus find u_0, the limit of u_ε. The homogenized process will be described by the homogenized quantity u_0,[9]-[11].

It should be noticed, however, that the homogenized equation could be different from the initial "heterogeneous" equation. For instance the Stokes equation in a heterogeneous porous medium made of obstacles and canals, with a periodic repartition, gives a homogenized equation which is the Darcy law. Even in the simplest case, the effective behavior is never straightforward [20].

To understand the difficulties we may look at the stationary heat equation:

$$-\nabla.(A^\varepsilon(x)\nabla u_\varepsilon) = f \qquad (3.1)$$

with $A^\varepsilon(x) = A(x, x/\varepsilon)$ a rapidly oscillating function. Although the effective equation is of the same type:

$$-\nabla.(A^h(x)\nabla u_0) = f \qquad (3.2)$$

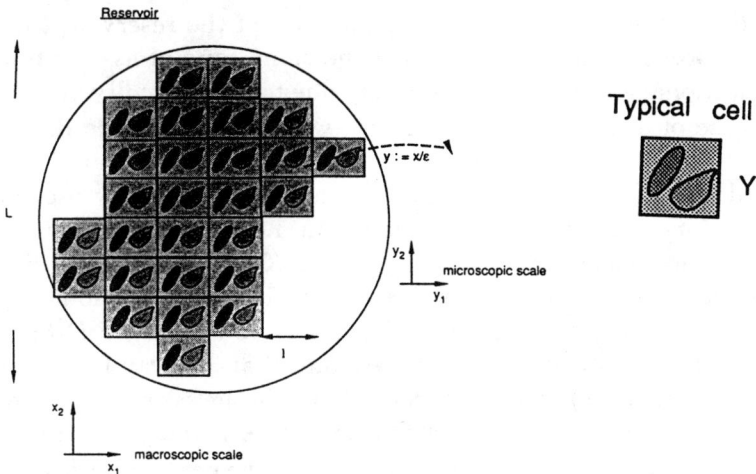

Figure 2. A heterogeneous porous medium: Macroscopic and microscopic scales.

where the effective tensor, A^h no longer depends on ε. It is proven, [9]-[11], that A^h, except in some very special cases, is neither the arithmetic mean nor the harmonic mean of A^ε. This is due to the fact that A^ε and ∇u_ε are two oscillatory quantities. Their product is not oscillating around the product of their arithmetic means. This can be seen in the example of $v^\varepsilon(x) = \sin(2\pi x/\varepsilon)$ oscillating around its mean 0 and $w^\varepsilon(x) = v^\varepsilon(x)v^\varepsilon(x)$ which does not oscillate around the product of the previous means (which would be 0) but oscillates around the arithmetic mean of w^ε which is 1/2.

These types of problems have been addressed by various authors in the field of petroleum engineering. Most of these works is based on "heuristic" considerations and take into account only incompressible one-phase flow in heterogeneous porous media. Moreover, the most common conclusion of these works is that the effective permeability lies between a harmonic and arithmetic average [21],[22], or is given by the geometric mean [23].

However, because homogenization is a rigorous mathematical method, some very precise hypotheses will be imposed. The first hypothesis is that we should know the mean around which coefficients are oscillating which in some sens a "weak" limit of the coefficients. The easiest way to have such an assumption is to assume that the coefficients are rapidly and quasi-periodically oscillating, i.e. depend on x and x/ε, with period ε. Any other assumption that gives the weak limits of the coefficients could have been used. The second hypothesis is that the region under consideration does not intersect any boundaries where boundary layers occur.

4. Theoretical Homogenization of Two-Phase Flow. In the framework described below we have considered two cases. The first one is the immiscible flow of two incompressible fluids, including the effects of capillary pressure and gravity and the second one is the case of miscible flow of two incompressible fluids including the effects of dispersion and gravity. In both cases we start from a description of the flow at the microscopic level of the heterogeneous porous medium (the small scale l is the scale of the heterogeneities of the porous medium), and we seek a homogenized description at the macroscopic level of the reservoir, (of lenght scale L). To get rigorous results we have used the same type of assumptions as above, and some additional assumptions essentially due to the difficulty of these problems.

In the case of a miscible displacement, we assume that the dispersion tensor D, depends only on the space variable x and on the concentration C. Moreover, it is assumed at fixed C that this tensor is rapidly periodically oscillating, taking different values in each type of porous medium, i.e. $D(x, C) = D_1(x)D_2(C)$. The porosity Φ and the absolute permeability tensor K of the rock take different values in each type of porous medium and then are rapidly quasi-periodically oscillating functions, $\Phi^\varepsilon(x) = \Phi(x, x/\varepsilon)$, $K^\varepsilon(x) = K(x, x/\varepsilon)$ and $D_1^\varepsilon(x) = D_1(x, x/\varepsilon)$. In the case of an immiscible displacement we assume that the capillary pressure curve $P_{cap}(x, S) = P_{cm}(x)p_c(x)$, does not depend on the heterogeneities and then it is assumed to be depending only on the space variable x and the saturation S.

In the first part we are assuming that the relative permeabilities $k_{ri}(S)$ and the function $p_c(S)$ have the same shape all over the domain of the reservoir Ω. Hence there is no jump in the saturation at the interfaces between heterogeneities. This would be rephrased by saying that we considered a field containing a single rock type. Under these assumptions we get the results below.

4.1 Theoretical Results on a Single Type Rock Model. From [1]-[6] the main results are:

1. The equations describing the homogenized flow are of the same type that the equations of the flow through the heterogeneous medium. There is an effective porosity Φ^h, effective absolute permeability tensor K^h and an effective dispersion tensor D_1^h which do not depend on ε.

2. Φ^h is the arithmetic average.

3. K^h and D_1^h are given by the same method as in the linear case, i.e. for instance K_{ij}^h is the arithmetic average of the sum of the coefficient K_{ij} and of a complementary term $\alpha_{ij} = \sum_{k=1}^n K_{ik}\frac{\partial w_j(y)}{\partial y_k}$, $1 \le i, j \le n$, where $w_j(y)$ is the periodic solution of a linear PDE in the microscopic variable $y = x/\varepsilon$ (see section 5).

Remark 1: The results given allow us to compute the homogenized flow in all cases of heterogeneities, with a quasi-periodic repartition, such as anisotropic media, various shapes of heterogeneities and several types of heterogeneities.

Remark 2: These results are consistent with the widely used rules of averaging when the flow is orthogonal or parallel to the stratas in a stratified porous medium.

Remark 3: In general cases, even with each type of porous media being isotropic, we have homogenized tensors wich are significantly anisotropic.

4.2 Theoretical Results of a Model with Different Rock Types. We may relax the assumption on the capillary pressure and the relative permeabilities curves, that is allow those curves to change in each type of porous medium. In each type j, there is p_c^j, Figure 3, and k_{ri}^j functions and hence there will be a saturation jump at the interface between each type of media [17]. For sake of simplicity, we will consider the case of a horizontal field Ω which contains only two different rock types, the subscript $j = 1, 2$ will refer to the jth type, and we will assume that $P_{cm}(x) = $ constant.

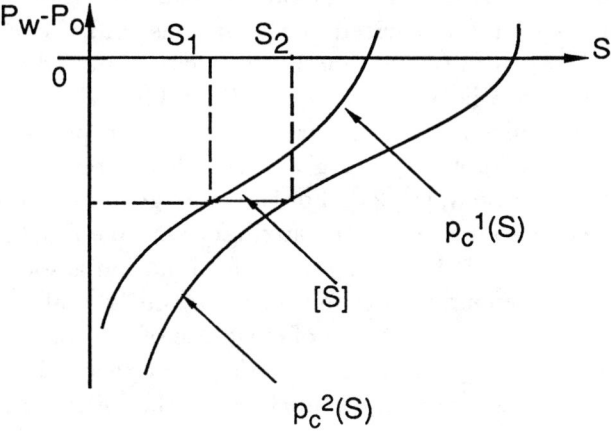

Figure 3. An example of the capillary pressure curves for the two media.

Let Ω_j, $j = 1, 2$, be the spatial domain of type j and Γ_{12} the interface between Ω_1 and Ω_2. There are in each Ω_j the same equations as (2.6)-(2.8), section 2:

$$q_t^j = -\lambda^j(S_j)K(x)\nabla P_j \tag{4.1}$$

$$\nabla.q_t^j = 0 \quad \text{in} \quad \Omega_j \times J \tag{4.2}$$

$$\Phi(x)\frac{\partial S_j}{\partial t} + \nabla.V_w^j = 0 \tag{4.3}$$

where

$$V_w^j = R^j + \lambda_w^j(S_j)q_t^j \tag{4.4}$$

and at the interface, the continuity of the physical quantities: P_w, P_o the water and oil pressures and the flux of each of the two fluids, gives the interface conditions:
1. continuity of the total flux:

$$[q_t.\nu] = 0 \tag{4.5}$$

2. continuity of the water flux:

$$[V_w.\nu] = 0 \tag{4.6}$$

3. continuity of the fluid pressure gives a jump on the "global" pressure, (2.5):

$$[P] = [\gamma(S)] \tag{4.7}$$

4. continuity of the capillary pressure gives a jump on the saturation S, Figure 3:

$$p_c^1(S_1) = p_c^2(S_2) \tag{4.8}$$

where ν is the outward normal to Γ_{12} and the brackets mean, as usual, the jump across Γ_{12}.

Essential difficulties of this problem are the nonlinearity arising in the interface condition (4.8), the discontinuity of the saturation and of the "global" pressure given by (4.7) and (4.8) on the interface and the supplementary coupling between the saturation equation (4.3) and the "global" pressure equation (4.1).

In a first step, we have linearized the equations and the capillary pressure curves, decoupled the system (cross mobility curves), assumed the "global" pressure jump at the interface is of order $O(\varepsilon)$. By taking $P_{cap}(S)$ as an auxiliary variable we find homogenized equations are of the same type as the original. But absolute permeability and porosity are given by other expressions than the standard ones obtained in section 3, [1],[24]. For the more general case with the coupled system (4.1)-(4.8), we introduce again the auxiliary variable $P_{cap}(S)$ which is now a nonlinear function. This leads to a variational formulation associated to the semi-discretization in time. Then a mixed form is used for the "global" pressure equation. Up to now we have a mathematical proof of existence of the solution of this system. The homogenization of this problem is currently underway. It appears that the homogenized system will not be of the same form as the initial one except for some special cases.

5. Numerical Simulations.

5.1 Numerical Computations of the Homogenized Tensor.
In general n dimensional cases, to obtain the homogenized tensor, we have to solve n so called local problems which are linear PDEs on a basic cell with periodic boundary conditions. The idea is to obtain the functions $w_j(y)$, $j = 1, ..., n$ (see section 4.1). Namely

$$\begin{cases} w_j(y) \quad \text{Y-periodic} \\ - \nabla.(K(y)\nabla w_j(y)) = \nabla.(K(y)e_j) \qquad \text{in} \quad Y \end{cases} \qquad (5.1)$$

where e_j is the jth standard basis vector.

Numerical computations to obtain the homogenized coefficients of the entire reservoir have been carried out via a finite element method. A numerical code [25] has been performed to solve problem (5.1) and compute the homogenized tensor obtained in section 4.1, for any heterogeneities in a cell. The results of the approximate solution of equation (5.1) are used below in several test problems.

5.2 Numerical Simulations of Two-Phase Flow in a Reservoir.
In order to test the validity of the theoretical results, we have run various simulations for incompressible two-phase flow. Comparisons have been done between heterogeneous and homogenized simulations for saturations and pressures, visualised by means of contour plots, cross-sections of surfaces, and through the production curves associated with the amount of resident fluid (oil) produced in the simulation. There are essentially two groups of simulations that have been performed:
1. Simulation on core samples and stratified quarter five-spot patterns have been performed using the BIDIMIX code (INRIA-ELF-IFP,[26]-[28]) at INRIA Rocquencourt France. These simulations include the testing of the influence of boundary layers and the ratio of the scale $\varepsilon = l/L$ on the accuracy of the homogenized model.

Various Peclet numbers and mobilities have been tested (see [1],[24]).

2. 2-D heterogeneities in a quarter five-spot pattern have been tested at the University of Wyoming in Laramie Wyoming and Utah State University in Logan Utah. The goal of these studies is to test by numerical simulations homogenized coefficients predicted by the theory above. As a mean of comparison, the arithmetic mean, harmonic average and the geometric mean have been used.

The following graphs show some of the results from the simulations done at the University of Wyoming and Utah State University using the SIM2P2D code,[29]-[31].

5.3 Presentation of Numerical Results. Numerical experiments to test the homogenization method described here have been performed on a variety of problems including two-phase miscible and immiscible displacement regimes. Also a variety of flow parameters including the viscosity ratio between the two fluids and the permeability ratio between different regions in the porous medium. The results will be presented below in four parts delineating four representative test problems. In all test problems below a quarter five-spot pattern was used on a unit square. The simulator used in obtaining the numerical results features the use of an IMPES solution of the pressure/velocity equation and saturation equation. The pressure and velocity equations are solved implicitly in the simulator to obtain approximations to the pressure and velocity at a new time level and then the saturation of the water phase is obtained via some explicit update given the total fluid velocity [30], [31]. The explicit update of the saturation equation for the test problems below was performed using one-point upwinding.

The simulations done in all cases had a fixed porosity, $\Phi = 0.2$, fixed injection and production rates at the wells, $q_0 = 0.25$ barrel/day, and the fuids were assumed to be incompressible. the relative permeabilities used were $k_{rw}(S) = S^2(3 - 2S)$, $k_{ro}(S) = 1 - k_{rw}(S)$. Finally, it should be noted that the absolute permeability tensor in all cases below was taken to be diagonal with components k_{xx}^j and k_{yy}^j, where j is an index that refers to different regions in the domain. If a cell has two regions of permeability then j would be defined for region I and II (see Figure 4).

5.3.1 Test Problem #1. The first test problem involved simulation of a quarter five-spot pattern with heterogeneities as shown in Figure 4. Figure (4b) shows a typical domain for the simulation where a repeated pattern of heterogeneous cells occurs. Figure (4a) shows a close up view of one of the heterogeneous cells. The shaded region in the interior of the cell is of low permeability, while the border is a region of higher permeability. For the simulations done for test problem #1 the relative dimensions for the two regions of differing permeabilities is shown in Figure (4a). Typical values chosen through out the numerical experiments for the ratio of the permeabilities in the two regions are 10:1, 100:1 and 1000:1.

For test problem #1, the permeability ratio was taken as 10:1. That is the two diagonal components are defined by $K_{xx}^I = K_{yy}^I = 10$ md $K_{xx}^{II} = K_{yy}^{II} = 1$ md. In this case the homogenized value of the permeablity is given by $K_{xx}^h = K_{yy}^h = 6.25$ md, while for the arithmetic mean = 7.75 md, the harmonic average = 3.07 md, and the geometric mean = 3.33 md. The viscosity ratio for this problem was taken

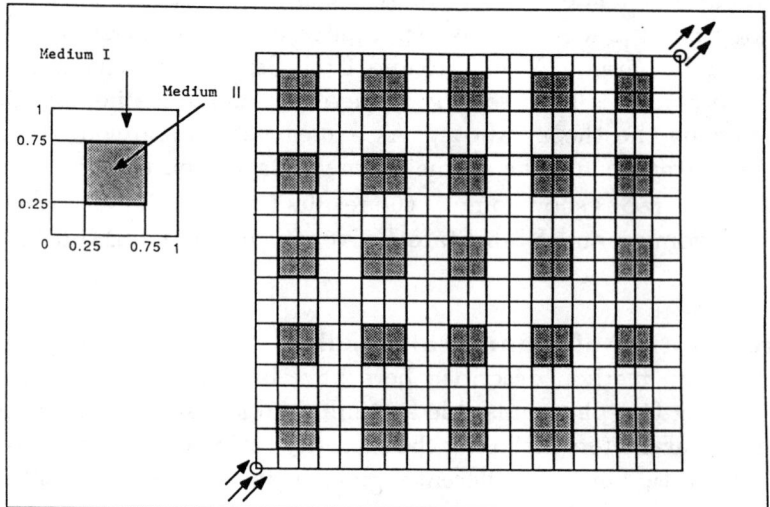

Figure 4. A typicall cell and a heterogeneous quarter five-spot 5 × 5 cells, 20 × 20 grid.

to be 1:1. The results of two different simulations are shown in Figures (5) through (9). The first simulation performed involved a 5 × 5 pattern of cells as shown in Figure (4b) (20 × 20 computational grid) and the second simulation involved 10 × 10 pattern of cells (40 × 40 computational grid) on the same domain.

Figure (5a) shows the saturation contours for the 5 × 5 cell pattern before break-through occurs at the production well. Figure (5b) shows the saturation contours at the same time on the homogenized medium. Figures (6a) and (6b) show the saturation contours after breakthrough has occured at the production well. Figures (7) and (8) show analogous results in the 10 × 10 cell pattern for the second simu-lation in test problem #1. Figure (9) shows the production curves measuring the amount of oil produced at the production well for the two simulations done under this test problem.

The results indicate reasonable agreement between the heterogeneous and ho-mogenized simulations. The production curves indicate that as the cell is made smaller respect to the domain lenght scale ($\varepsilon \longrightarrow 0$) the approximation is better as one would expect.

5.3.2 Test Problem #2. Under test problem #2, two simulations were again done. The simulation involved the same cell patterns as in test problem #1. In this case, the viscosity ratio of the fluids was chosen to be 50:1. That is the resident oil phase was 50 time as viscous as the water phase being injected. Note that the homogenized value of the permeability will remain the same in this test problem.

The results for the second test problem are shown in Figures (10) through (14). The graphics from the simulations are shown in the same order as in test problem #1. The results of the simulations again show good agreement between the heterogeneous simulations and the homogenized simulations. Again the production curves show similar improvement as the ratio of the cells is decreased relative to

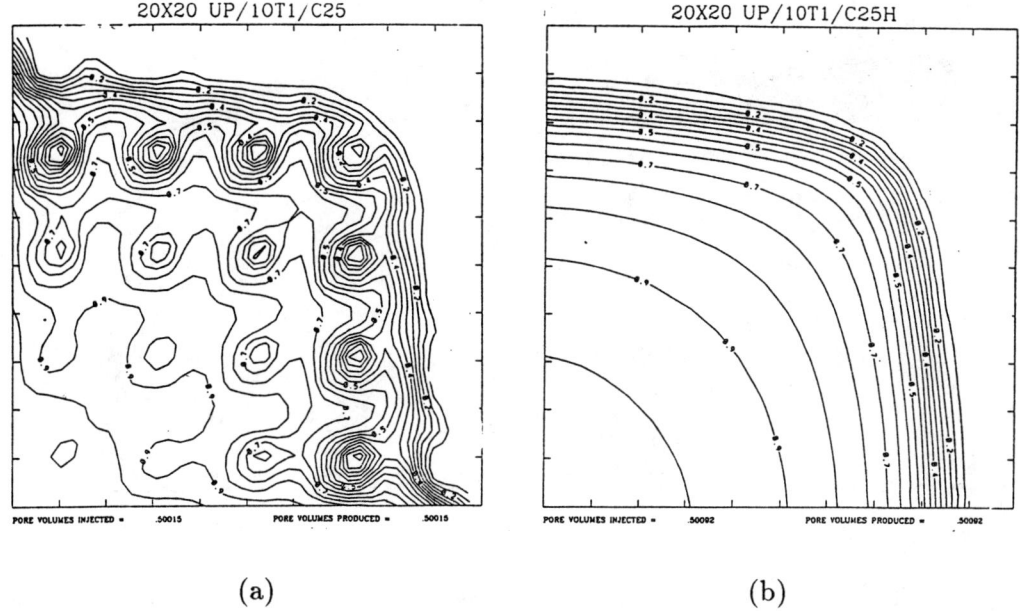

Figure 5. The saturations contours before breakthrough at the production well for (a) the heterogeneous simulation and (b) for the homogenized simulation with a 5 × 5 **cell** pattern for test problem #1.

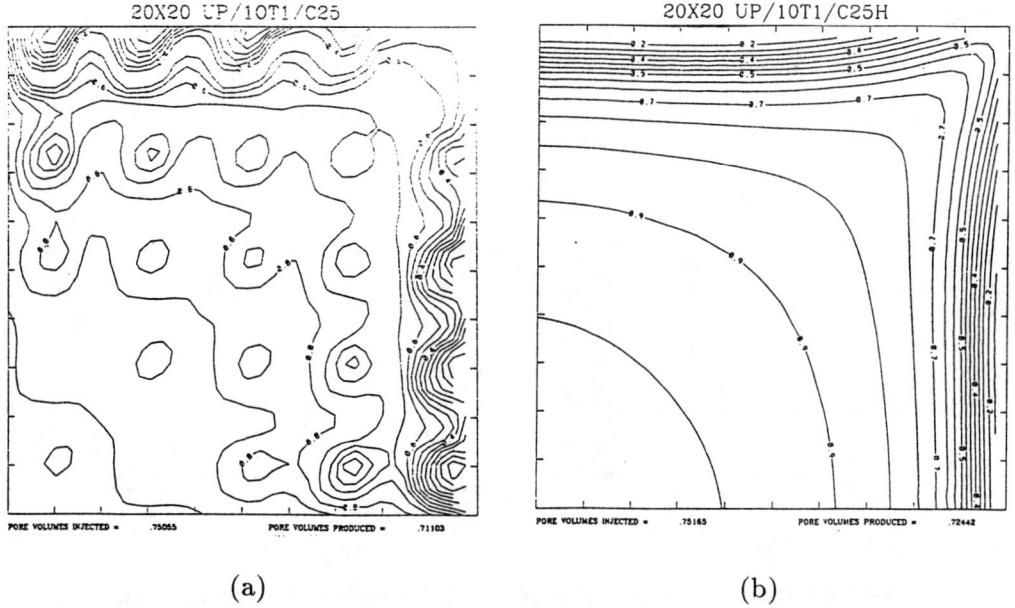

Figure 6. The saturations contours after breakthrough at the production well for (a) the heterogeneous simulation and (b) for the homogenized simulation with a 5 × 5 **cell** pattern for test problem #1.

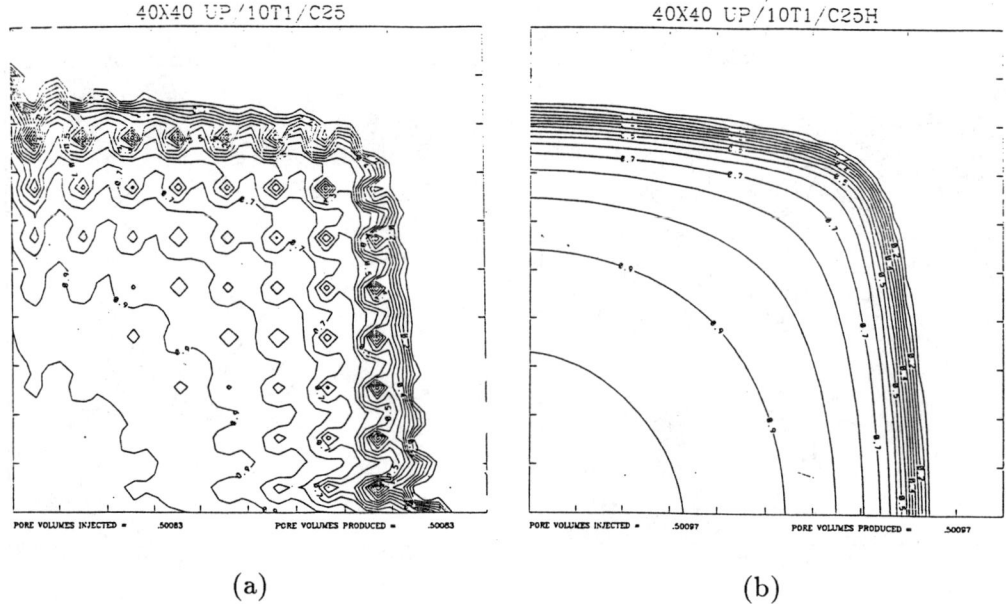

(a)   (b)

Figure 7. The saturations contours before breakthrough at the production well for (a) the heterogeneous simulation and (b) for the homogenized simulation with a 10×10 **cell** pattern for test problem #1.

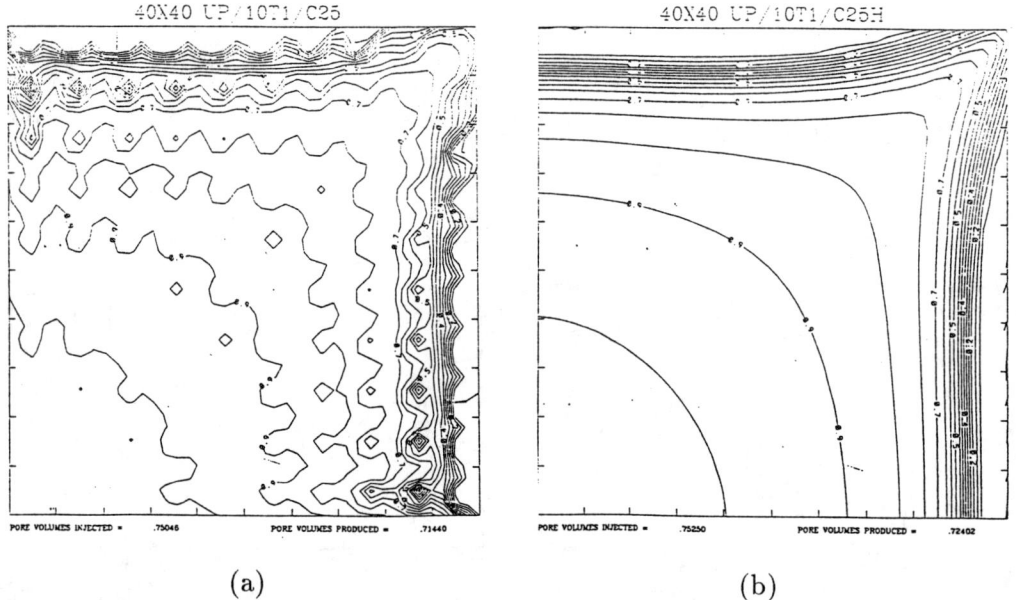

(a)   (b)

Figure 8. The saturations contours after breakthrough at the production well for (a) the heterogeneous simulation and (b) for the homogenized simulation with a 10×10 **cell** pattern for test problem #1.

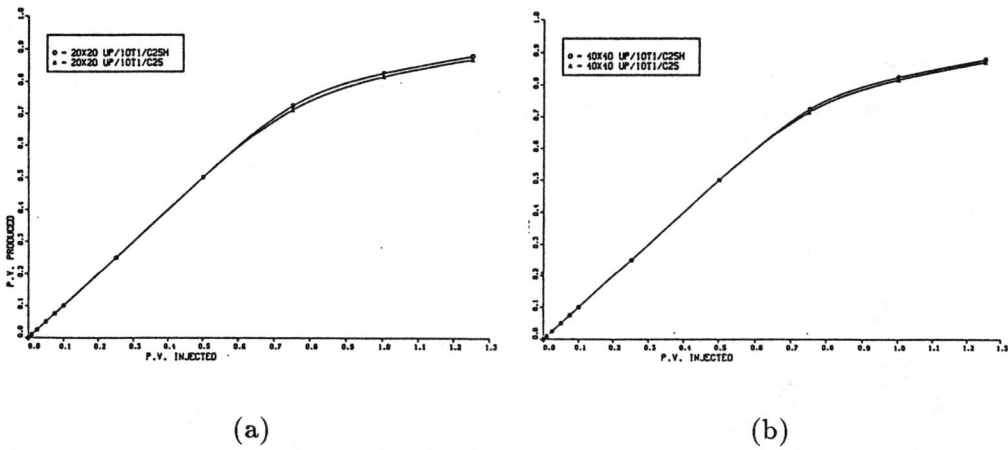

(a)   (b)

Figure 9. The production curves for the heterogeneous and homogenized simulations are plotted for (a) a 5×5 **cell** pattern and (b) a 10×10 **cell** pattern for test problem #1.

the length scale of the whole domain as in test problem #1.

5.3.3 Test Problem #3. The third test problem uses the same pattern of heterogeneities as the first two test problems, but changes the permeability ratio to 100:1. The associated homogenized value of the permeability is $K_{xx}^h = K_{yy}^h = 59.2$ md. The viscosity ratio is 50:1 as in test problem #2. It should be noted at this point that the viscosity ratio being 50:1 only corresponds to a mobility ratio of about 0.4 at the front for the cubic relative permeabilities curves used in all the problems. Thus the fingering effect will not be as great for this problem. For this test problem only one set of simulations is presented. The case of a 5×5 pattern of cells is considered in the figures.

5.3.4 Test Problem #4. In test problem #4, The pattern was changed slightly and is shown in Figure (17). In this case the cell pattern is changed only in the dimensions of the regions containing the two media. The width of the high permeability region is only one tenth the total well width limiting the flow path to the previous test problems. The viscosity ratio was fixed at 1:1 and the permeability ratio is fixed at 10:1 as in the first test problem. The homogenized permeability value is $K_{xx}^h = K_{yy}^h = 3.106$ md. The results below were compared to simulations performed using the arithmetic mean = 4.24 md, the harmonic average = 1.48 md, and the geometric mean = 3.33 md. We should note that in this case the values for the homogenized and geometric mean coefficient values are very close.

The results shown in Figures (18) through (20) show the same type of agree-

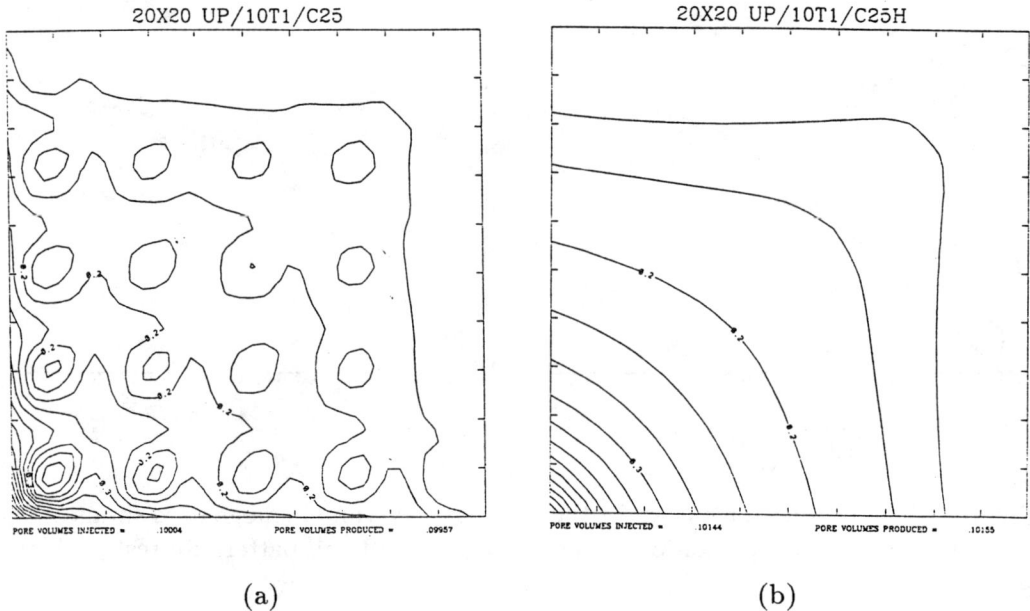

(a)   (b)

Figure 10. The saturations contours before breakthrough at the production well
for (a) the heterogeneous simulation and (b) for the homogenized simulation with
a 5 × 5 **cell** pattern for test problem #2.

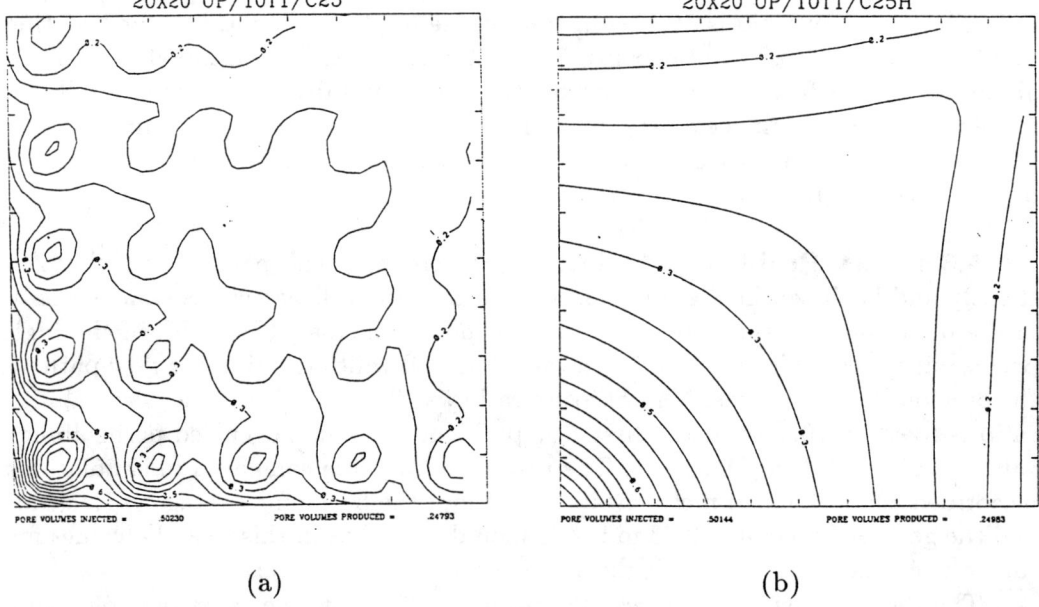

(a)   (b)

Figure 11. The saturations contours after breakthrough at the production well for
(a) the heterogeneous simulation and (b) for the homogenized simulation with a
5 × 5 **cell** pattern for test problem #2.

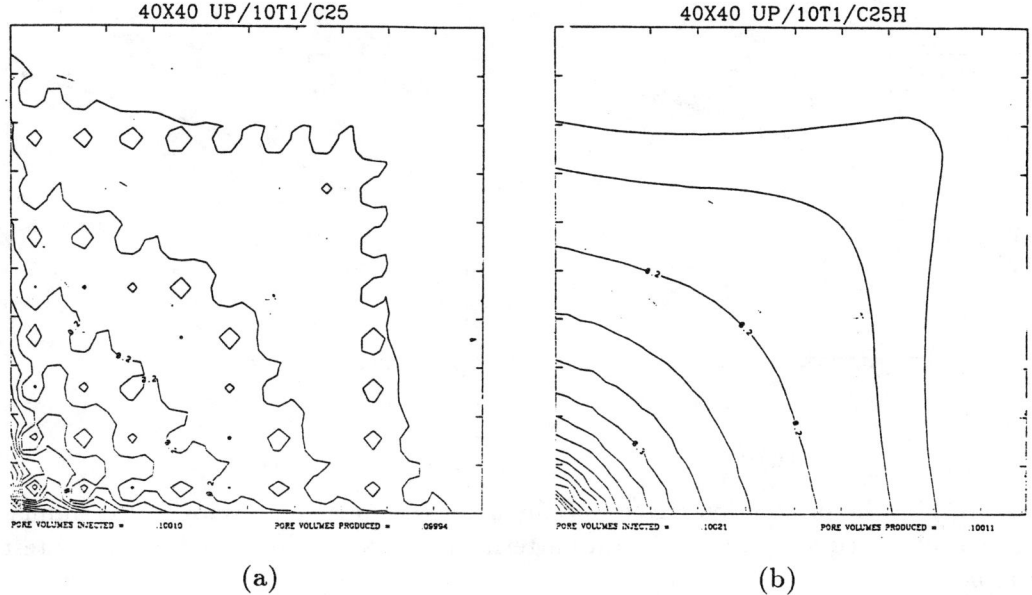

Figure 12. The saturations contours before breakthrough at the production well for (a) the heterogeneous simulation and (b) for the homogenized simulation with a 10 × 10 **cell** pattern for test problem #2.

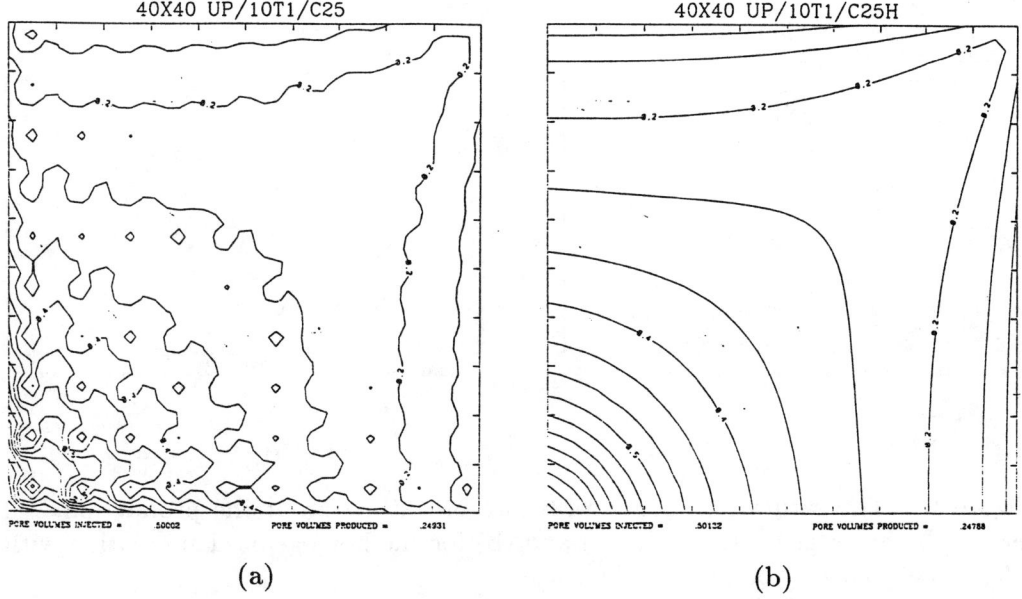

Figure 13. The saturations contours after breakthrough at the production well for (a) the heterogeneous simulation and (b) for the homogenized simulation with a 10 × 10 **cell** pattern for test problem #2.

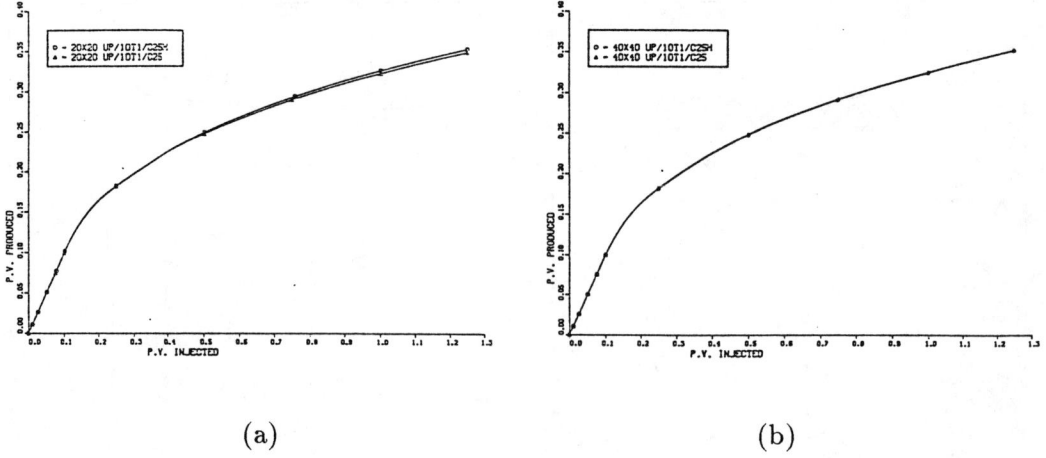

(a)   (b)

Figure 14. The production curves for the heterogeneous and homogenized simulations are plotted for (a) a 5×5 **cell** pattern and (b) a 10×10 **cell** pattern for test problem #2.

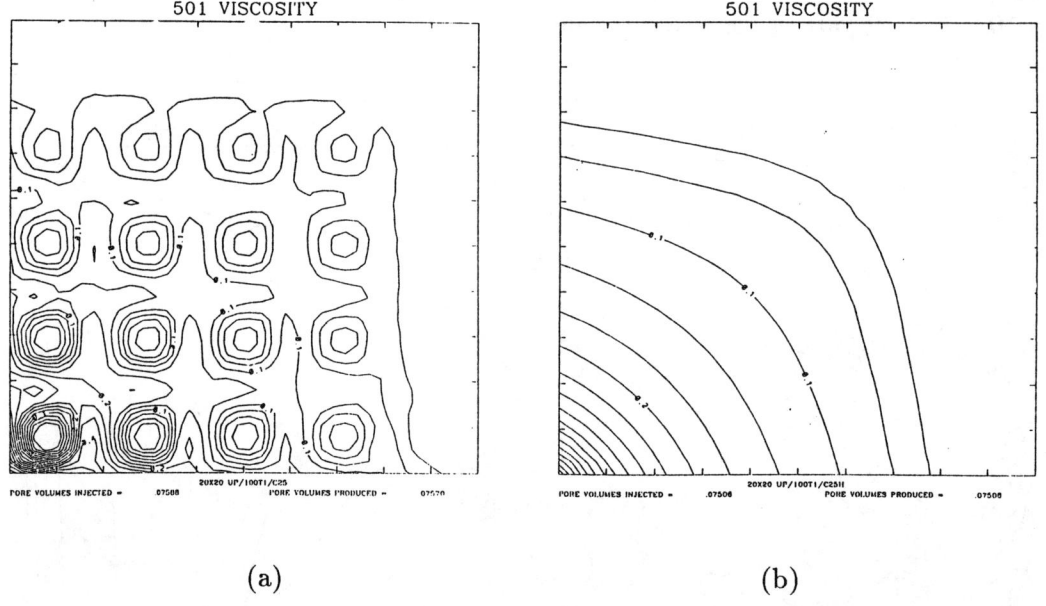

(a)   (b)

Figure 15. The saturations contours before breakthrough at the production well for (a) the heterogeneous simulation and (b) for the homogenized simulation with a 5×5 **cell** pattern for test problem #3.

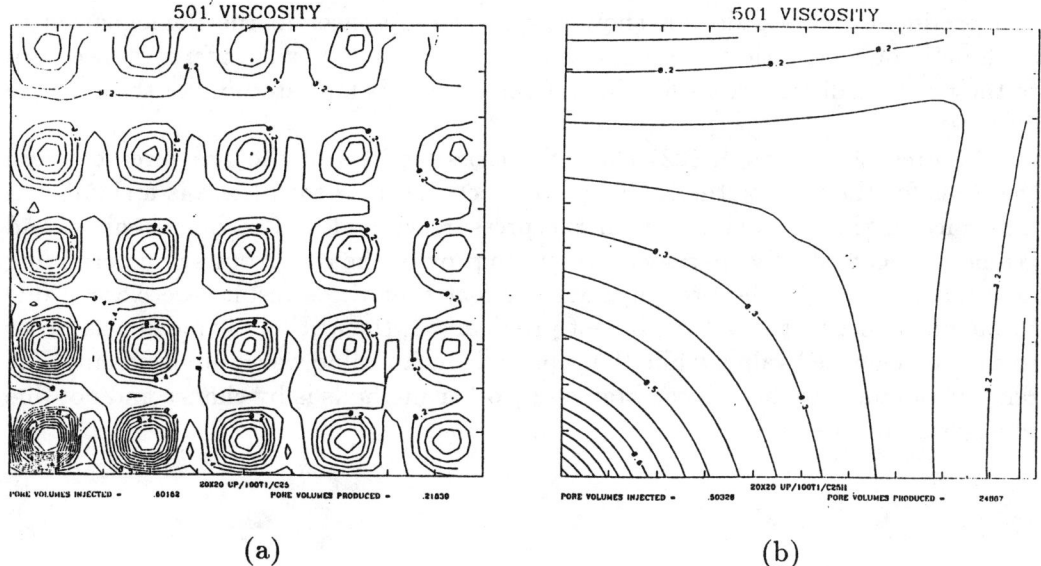

(a)   (b)

Figure 16. The saturations contours after breakthrough at the production well for (a) the heterogeneous simulation and (b) for the homogenized simulation with a 5 × 5 **cell** pattern for test problem #3.

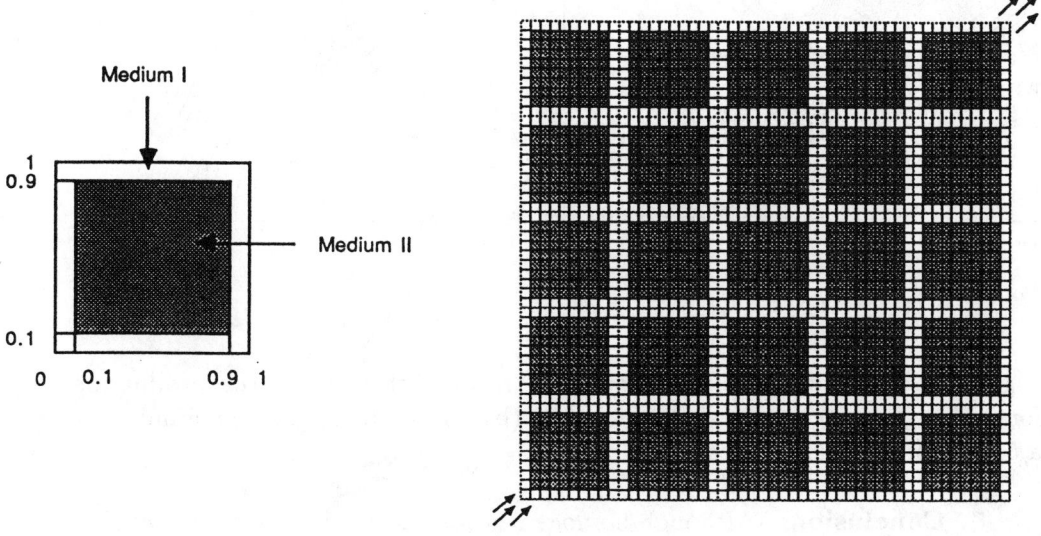

Figure 17. Depiction of a cell and a heterogeneous quarter five-spot 5 × 5 cells, 50 × 50 grid.

ment as in the first two test problem results presented. The saturation curves and production curves do not show any great differences in the homogenized or arithmetic or geometric mean simulations. To see significant differences we turn to the pressure distributions for the different methods for determining the effective parameters.

Figures (21) through (22) show the cross-sections of the pressures between the wells for the four methods. The cross-sections show that there is a significant difference in the approximation to the pressure variable. The homogenized and geometric mean effective parameters seem to provide the best results in determining an approximation to the pressure, but the fact that these results are close is due to the particular test problem. In test problem #1, the arithmetic mean was closer to the homogenized value, while the geometric mean value was quite different. The relative closeness of the values obtained by other methods is by chance and not due to any rigorous theory.

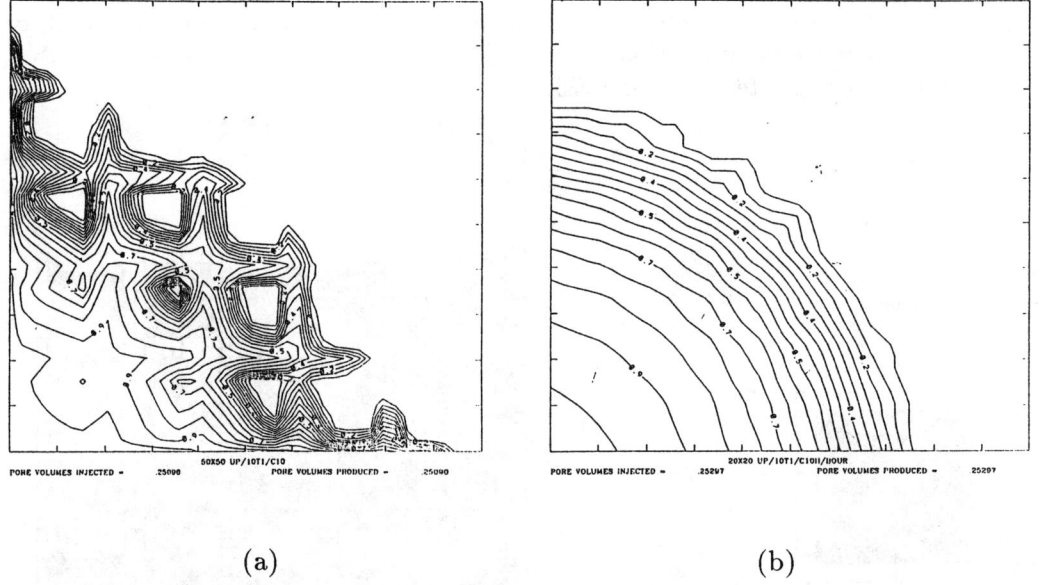

(a)   (b)

Figure 18. The saturations contours before breakthrough at the production well for (a) the heterogeneous simulation and (b) for the homogenized simulation with a 5 × 5 **cell** pattern for test problem #4.

6. Conclusions. Although homogenization theory does not take into account boundary layers, all our test cases have been run on a quarter five-spot patterns which have real boundary layers near the boundaries of the domain and near the wells. These boundary layers, however, have not effected the results of the simulations adversely.

In the same way one main assumption is that the size of the heterogeneities is on the same order as the cell size ε. But in test problem #4, the cell size is

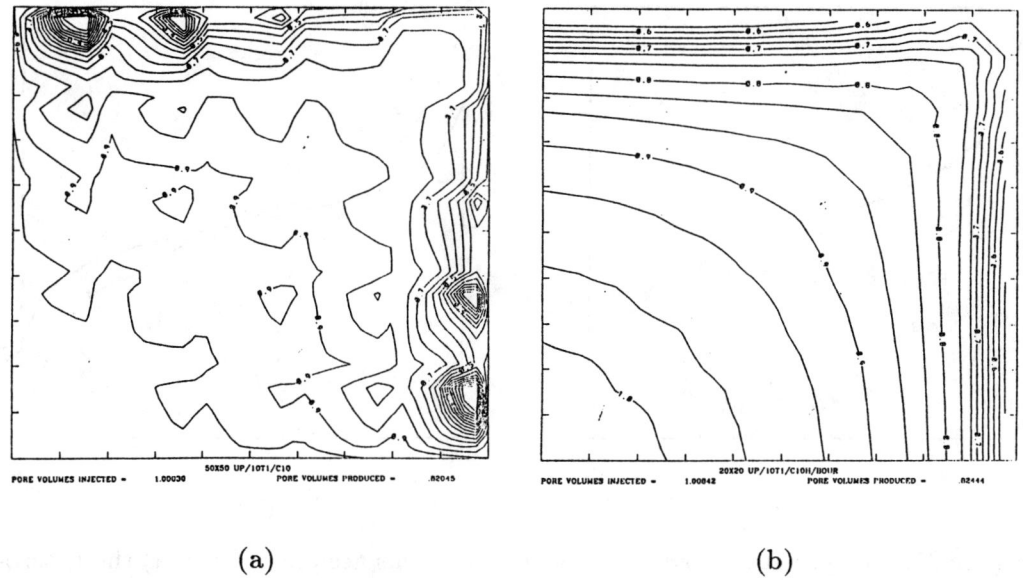

(a)   (b)

Figure 19. The saturations contours after breakthrough at the production well for (a) the heterogeneous simulation and (b) for the homogenized simulation with a 5 × 5 **cell** pattern for test problem #4.

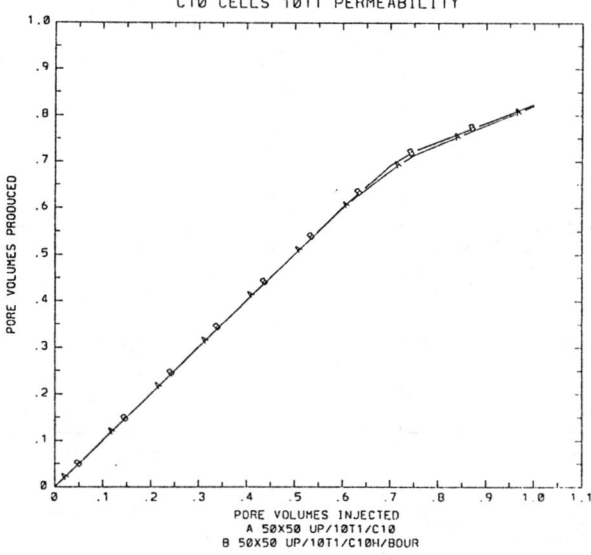

Figure 20. The production curves for the heterogeneous and homogenized simulations are plotted for a 5 × 5 **cell** pattern for test problem #4.

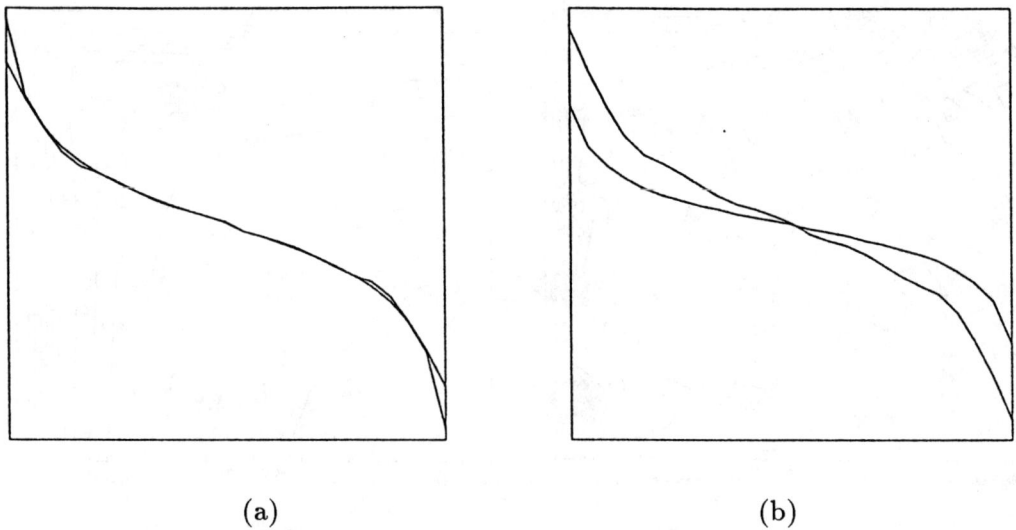

(a)   (b)

Figure 21. The pressure cross-sections between wells are plotted for (a) the hetero-geneous and **homogenized** simulations and (b) for the heterogeneous and **arith-metic** average simulations for a 5×5 **cell** pattern for test problem #4.

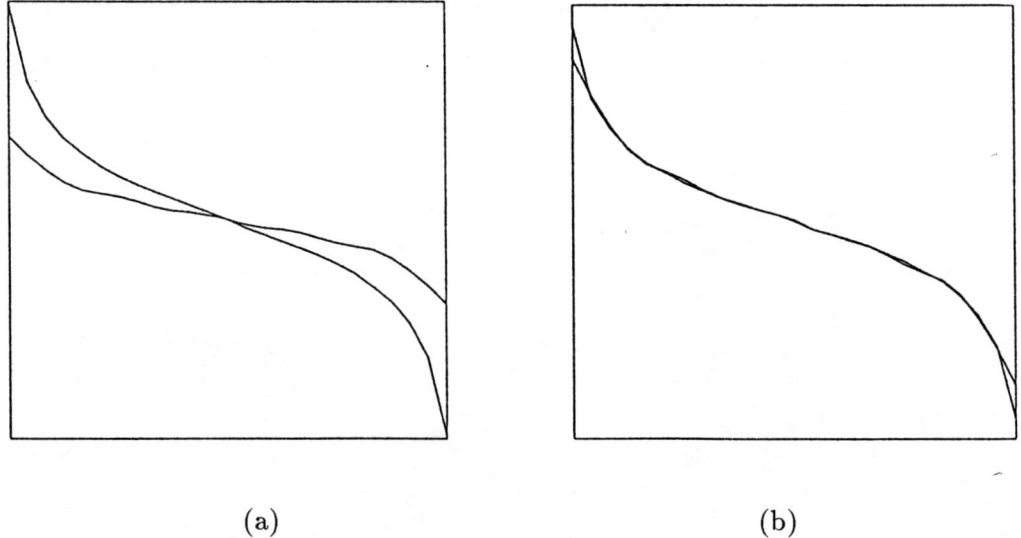

(a)   (b)

Figure 22. The pressure cross-sections between wells are plotted for (a) the het-erogeneous and **harmonic** average simulations and (b) for the heterogeneous and **geometric** average simulations for a 5×5 **cell** pattern for test problem #4.

$\varepsilon = 1/5$ and the "fissures" have dimension $1/10$ of the cell size, i.e. $1/50$. Even though the heterogeneity is relatively small compared with ε one should notice that the homogenized results are in good agreement with the heterogeneous results.

Because all of the simulations have been done with a constant injection rate, the saturation contours are not very sensitive to any change in the absolute permeability, but the pressure field does change dramatically. Thus the trust test in the results is in how well the homogenized pressure fields match the heterogeneous pressure field. In the results presented the best match of the pressure fields is obtained by using the homogenization theory presented in this paper. The other averaging methods produce inaccurate pressure fields.

Finally, taking into account previous numerical simulations reported in [1] and [24], on core samples and stratas, including various Peclet numbers and strong boundary effects one can conclude that the homogenization technique presented here is a robust method.

Acknowledgements: We would like to thank the Institute for Scientific Computation at the University of Wyoming and Chevron Oil Field Research Company for allowing us to use their computational facilities in this work.

REFERENCES

1. B. AMAZIANE, *Application des techniques d'homogénéisation aux écoulements diphasiques incompressibles en milieu poreux*, PhD Thesis, University of Lyon I, 1988. (In French.)

2. A. BOURGEAT, *Homogenization method applied to the behavior of a naturally fissured reservoir*, in Mathematical Method in Energy Research, K. J. Gross, ed., SIAM, 1984, pp. 181–193.

3. A. BOURGEAT, *Homogenized behavior of diphasic flow in a naturally fissured reservoir with uniform fractures distribution*, Comput. Meth. Appl. Mech. Eng., 47 (1984), pp. 205–217.

4. A. BOURGEAT, *Nonlinear homogenization of two-phase flow equations*, in Physical Mathematics and Nonlinear Differential Equations, J. H. Lightbourne and S. M. Rankin, eds., Lecture Notes in Pure and Applied Mathematics, Marcel Dekker, New-York, 1985, pp. 207–212.

5. A. BOURGEAT, *Global behavior of two-phase flow in inhomogeneous media*, in 3rd International Conference on Boundary Layers, Boole Press Pub., Dublin, 1985, pp. 157–160.

6. A. BOURGEAT, *Homogenization of two-phase flow equations*, in Nonlinear Functional Analysis and its Applications, Proceedings of Symposia in Pure Mathematics, Volume 45–Part 1, F. E. Browder, ed., AMS, 1986, pp. 157–163.

7. Y. AMIRAT, K. HAMDACHE, A. ZIANI, *Homogénéisation d'un modèle d'écoulements miscibles en milieu poreux*, INRIA, 802, 1988. (In French.)

8. Y. AMIRAT, K. HAMDACHE, A. ZIANI, *Homogénéisation d'équations hyperboliques du premier ordre: application aux milieux poreux*, INRIA, 803, 1988. (In French.)

9. A. BENSOUSSAN, J. L. LIONS, G. PAPANICOLAOU, *Asymptotic Analysis*

for Periodic Structures, North-Holland, Amsterdam, 1978.

10. E. SANCHEZ-PALENCIA, *Non-Homogeneous Media and Vibration Theory*, Lecture Notes in Physics, 127, Springer-Verlag, 1980.

11. F. MURAT, *H-Convergence*, Séminaire d'Analyse Fonctionnelle et Numérique, Université d'Alger, Multigraphié, 1978. (In French.)

12. G. PAPANICOLAOU, S. VARADAN, *Diffusion in region with many small holes*, Lecture Notes Control and Information, 75, Springer-Verlag, 1980, pp. 190–206.

13. A. BOURGEAT, M. QUINTARD, S. WHITAKER, *Eléments de comparaison entre la méthode d'homogénéisation et la méthode de prise de moyenne avec fermeture*, C.R.A.S. Paris, T. 360, Série II, (1988), pp. 463–466. (In French.)

14. H. C. CHANG, *Multiscale analysis of effective transport in periodic heterogeneous media*, Chem. Eng. Comm., 15 (1982), pp. 83–91.

15. R. E. EWING, *Problems arising in the modeling of processes for hydrocarbon recovery*, in The Mathematics of Reservoir Simulation, R. E. Ewing, ed., Frontiers in Applied Mathematics, Volume 1, SIAM, 1983, pp. 3–34.

16. D. W. PEACEMAN, *Fundamentals of Numerical Reservoir Engineering*, Elsevier Scientific Publishing, New-York, 1977.

17. G. CHAVENT, J. JAFFRE, *Mathematical Models and Finite Elements for Reservoir Simulation*, North-Holland, Amsterdam, 1986.

18. G. CHAVENT, *A new formulation of diphasic incompressible flows in porous media*, in Lecture Notes in Mathematics, 503, Springer-Verlag, 1976, pp. 258–270.

19. S. N. KRUZKOV, S. M. SUKORJANSKII, *Boundary value problems for systems of two-phase porous flow type: statement of the problems, questions of solvability, justification of approximate methods*, Math. USSR Sbornik, 33 (1977), pp. 62–80.

20. L. TARTAR, *Incompressible fluid flow in a porous medium. Convergence of the homogenization process*, Appendix to book by E. Sanchez-Palencia above.

21. W. T. CARDWELL, R. L. PARSONS, *Average permeabilities of heterogeneous oil sands*, Trans. AIME, 160 (1945), pp. 34–47.

22. G. MATHERON, *Eléments pour une Théorie des Milieux Poreux*, Masson, Paris, 1976. (In French.)

23. J. E. WARREN, F. F. SKIBA, *Macroscopic dispersion*, Soc. Pet. Eng. J., (1964), pp. 215–230.

24. B. AMAZIANE, A. BOURGEAT, *Effective behavior of two-phase flow in heterogeneous reservoir*, in Numerical Simulation in Oil Recovery, The IMA Volumes in Mathematics and its Applications, M. F. Wheeler, ed., Volume 11, Springer-Verlag, 1988, pp. 1–22.

25. B. AMAZIANE, T. DUMONT, *Calcul de coefficients homogénéisés: Implémentation dans MODULEF et résultats numériques*, Publications du Laboratoire d'Analyse Numérique, URA-CNRS 740, Lyon & Saint-Etienne, 61 (1987). (In French.)

26. G. CHAVENT, G. COHEN, J. JAFFRE, *Discontinuous upwinding and mixed finite elements for two-phase flows in reservoir simulation*, Comput. Meth. Appl. Mech. Eng., 47 (1984), pp. 93–118.

27. G. CHAVENT, B. COCKBURN, G. COHEN, J. JAFFRE, *Une méthode d'éléments finis pour la simulation dans un réservoir de déplacements bidimensionnels*

d'huile par de l'eau, INRIA, 353, (1985). (In French.)

28. G. CHAVENT, J. JAFFRE, R. EYMARD, D. GUERILLOT, L. WEILL, *Discontinuous and mixed finite elements for two-phase incompressible flow.*, In 1987 SPE Reservoir Simulation Symposium, SPE, San Antonio, Texas, 1987. SPE Paper 16018.

29. M. B. ALLEN, R. E. EWING, J. V. KOEBBE, *Mixed finite element methods for computing groudwater velocities*, in International Conference Series on Advances in Numerical Methods in Engineering Theory and Applications, O. C. Zienkiewicz et al., eds., John Wiley & Sons, Inc., 1985, pp. 195–207.

30. R. E. EWING, M. F. WHEELER, *Computational aspects of mixed finite element methods*, in Numerical Methods for Scientific Computing, R. S. Stepleman, ed., North-Holland Publishing Co., 1983, pp. 163–172.

31. J. V. KOEBBE, *Numerical schemes for the immiscible displacement equations using a general polynomial framework for the saturation equation and mixed finite element methods for the pressure*, PhD Thesis, University of Wyoming, 1988.

CHAPTER 2

Numerical Simulation of Laboratory Displacement Experiments by a Petrov-Galerkin Method and a Pseudospectral Fourier Method

Ø. Bøe*
S. O. Hestholm*
A. Kamel*

Abstract. Numerical simulation of displacement experiments on a laboratory scale is an important tool for generating relative permeability curves. These curves are required as input for large scale simulations of multiphase flow in petroleum reservoirs. We study the displacement of a *nonwetting* fluid (oil) by a *wetting* fluid (water) and of a wetting fluid (water) by a nonwetting fluid (oil) in a core sample. Such experiments are named *imbibition* and *drainage*, respectively. We model these processes by one dimensional incompressible flow. In this work, two methods for numerical simulation of displacement experiments are studied and compared. The first is a *Petrov-Galerkin method* with linear trial functions and quadratic, asymmetric test functions. The second is the *pseudospectral Fourier method*. For displacement experiments, correct modelling of the boundary conditions is important, especially for small core samples. Comparison of the treatment of boundary conditions, computational cost and accuracy are given.

The outline of this paper is as follows: In part 1 we present the mathematical model. The Petrov-Galerkin and pseudospectral schemes are described in part 2. Numerical examples and a comparison of the two schemes are presented in part 3.

1 MATHEMATICAL MODEL. Consider a cylindrical core sample of length L which is placed in the gravitational field. Let the x-axis be parallel to the length axis of the cylinder. We consider the following problem: At one end of the cylinder (the inlet) we inject a wetting (nonwetting) fluid which displaces a nonwetting (wetting) fluid. If the displaced fluid is a nonwetting (resp. wetting) fluid, the process is called imbibition (resp. drainage). We consider the homogeneous incompressible case, so the rock dependent absolute permeability k, the porosity ϕ, the densities ρ_{nw} and ρ_w, and the viscosities μ_{nw} and μ_w are constants. g is the component of the gravitational acceleration vector in the positive x-direction. We assume that the flow is one dimensional, i.e in every circular slice orthogonal to the length axis of the cylinder all pressures and saturations are constants.

---

*IBM Bergen Scientific Centre, Thormøhlensgt.55, N-5008 Bergen, Norway.

Darcy's law and the equations of mass conservation for each fluid read:

$$u_w = -\frac{k k_{r_w}}{\mu_w}\left(\frac{\partial p_w}{\partial x} - \rho_w \vec{g}\right) \quad , \text{Darcy's law for wetting phase,} \qquad (1.1)$$

$$u_{nw} = -\frac{k k_{r_{nw}}}{\mu_o}\left(\frac{\partial p_{nw}}{\partial x} - \rho_{nw} \vec{g}\right) , \text{Darcy's law for nonwetting phase.} \quad (1.2)$$

$$\frac{\partial u_w}{\partial x} + \phi \frac{\partial S_w}{\partial t} = 0 \, , \text{mass conservation of wetting phase,} \qquad (1.3)$$

$$\frac{\partial u_{nw}}{\partial x} + \phi \frac{\partial S_{nw}}{\partial t} = 0 \, , \text{mass conservation of nonwetting phase.} \qquad (1.4)$$

k_{r_w} and $k_{r_{nw}}$ are *the relative permeabilities* for the wetting and nonwetting phase. They are assumed to be known monotonic increasing functions of the phase saturations, confer Figure 1.1. k_{r_w} vanishes at $S_w = S_{wc}$ (irreducible wetting phase saturation) and $k_{r_{nw}}$ vanishes at $S_{nw} = S_{nwr}$ (residual nonwetting phase saturation). For phase saturations below these values the phase is *immobile*. The *Darcy velocities*, u_w and u_{nw} , the phase pressures, p_w and p_{nw}, and the saturations, S_w and S_{nw} make a total number of six unknowns. The two other equations needed to close the system are the *the capillary pressure relation* $P_c = p_{nw} - p_w$ (Figure 1.1), and the volume balance of the saturations:

$$P_c(w) = p_{nw} - p_w \geq 0 \, , \text{capillary pressure,} \qquad (1.5)$$

$$S_w + S_{nw} = 1 \qquad , \text{volume balance.} \qquad (1.6)$$

For the capillary pressure curve we assume:

$$\lim_{S_w \to S_{wc}} \frac{dP_c}{dS_w} \to -M_1 \, , 0 < M_1 \leq \infty, \qquad (1.7)$$

$$\lim_{S_w \to 1 - S_{nwr}} \frac{dP_c}{dS_w} \to -M_2 \, , 0 \leq M_2 \leq \infty. \qquad (1.8)$$

The sum of the Darcy velocities u_w and u_{nw} is the total Darcy velocity:

$$u = u_w + u_{nw}, \qquad \text{total Darcy velocity.} \qquad (1.9)$$

For simplicity, we assume constant rate of injection. It follows from the definition of the total Darcy velocity u and the adding of the mass conservation equations that $u = $ constant.

We introduce dimensionless variables:

$$\tau = \frac{ut}{\phi L} \, , \text{number of injected pore volumes,} \qquad (1.10)$$

$$\eta = \frac{x}{L} \quad , \text{normalized plug length.} \qquad (1.11)$$

From now on we take $S = S_w$ to be the saturation in the wetting phase, and let $f = u_w$ be the Darcy velocity of the wetting phase.

By using (1.3)-(1.6) the one dimensional saturation equation is derived [1]:

$$\frac{\partial S}{\partial \tau} + \frac{\partial f}{\partial \eta} = 0, \qquad (1.12)$$

$$f = f_c(S) + f_d(S), \qquad \text{Darcy velocity of wetting fluid,} \qquad (1.13)$$

$$f_c = A(S)u+$$
$$\qquad + kH(S)(\rho_w - \rho_{nw})g \qquad \text{convective part,} \qquad (1.14)$$

$$f_d = \frac{kH(S)}{L}\frac{\partial P_c(S)}{\partial \eta}, \qquad \text{diffusive part,} \qquad (1.15)$$

$$A = \frac{\lambda_w(S)}{\lambda_w(S) + \lambda_{nw}(S)}, \qquad \text{mobility fraction,} \qquad (1.16)$$

$$H = \frac{\lambda_w(S) \cdot \lambda_{nw}(S)}{\lambda_w(S) + \lambda_{nw}(S)}, \qquad \text{harmonic average of mobilities,} \qquad (1.17)$$

$$\lambda_l(S) = \frac{k_{r_l}(S)}{\mu_l}, \qquad \text{mobility phase l; l=w,nw.} \qquad (1.18)$$

The saturation equation is a second order *quasilinear partial differential equation of parabolic type* as can be seen by substituting the definition of f given by (1.13)-(1.18) into (1.12).

The mobility fraction $A(S)$ is a monotonic increasing S-shaped function of S, while the harmonic average $H(S)$ is a bell shaped function, confer Figure 1.1. Both functions vanish at $S = S_{wc}$, and $H(S)$ also vanishes at $S = 1 - S_{nwr}$. This is a *singular* incident and the saturation equation degenerates into a *hyperbolic* equation.

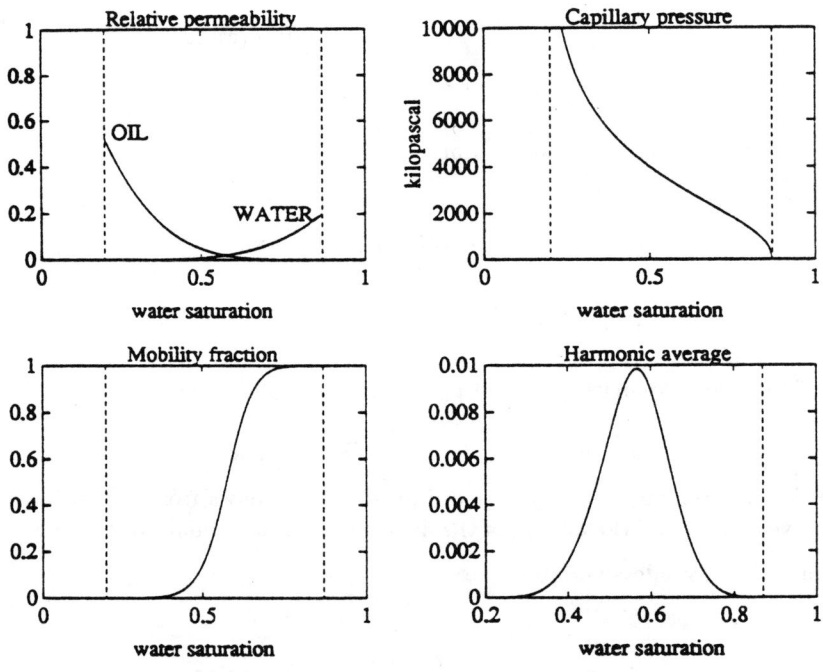

Figure 1.1 Saturation dependent parameter functions.

Boundary and initial conditions. At the inlet the total flow rate of the displacing fluid is equal to the total Darcy flow rate. In the imbibition case the capillary pressure leads to the so called *end effect* at the outlet. $P_c = 0$ for $S = 1 - S_{nwr}$ where S_{nwr} is the residual saturation in the nonwetting fluid. Before mobile water reaches the outlet end, oil is flowing from that end. Because of continuity of pressures, water cannot emerge (break through) from the outlet before the wetting saturation reaches $S = 1 - S_{nwr}$ and the phase pressures are equal.

Let τ_b denote the time for water break through. Note that for $\tau > \tau_b$ the flow of the nonwetting phase at the outlet may be different from zero even though $\lambda_{nw}|_{1-S_{nw}} = 0$. This happens because the pressure gradient $\partial p_{nw}/\partial \eta$ may be locally infinite at $\eta = 1$, giving a finite Darcy velocity $u_{nw}|_{\eta=1}$ for the nonwetting phase, according to Darcy's law, (1.4). After breakthrough the Dirichlet condition $S|_{\eta=1} = 1 - S_{nwr}$ is given at the outlet. The boundary conditions are given by:

Imbibition:
$$f|_{\eta=0} = u, \quad \text{all } \tau, \tag{1.19}$$

$$f|_{\eta=1} = 0, \quad \tau < \tau_b \text{ (as long as } S < 1 - S_{nwr}), \tag{1.20}$$

$$S|_{\eta=1} = 1 - S_{nwr} \; \tau \geq \tau_b. \tag{1.21}$$

Drainage:
$$f|_{\eta=0} = 0, \quad \text{all } \tau, \tag{1.22}$$

$$S|_{\eta=1} = 1 - S_{nwr} \quad \text{all } \tau. \tag{1.23}$$

The condition (1.20) at the outlet boundary gives (by using (1.13)-(1.18)):

$$u = -k\lambda_w[\frac{1}{L}P_c'(S)\frac{\partial S}{\partial \eta} + (\rho_w - \rho_o)g]|_{\eta=1}. \tag{1.24}$$

We now take the limit as $S \to 1 - S_{nwr}$ and assume M_2 in (1.8) to be finite:

$$u = \lim_{S_w \to 1-S_{nwr}} \left(k\lambda_w[\frac{1}{L}P_c'(S)\frac{\partial S}{\partial \eta} + (\rho_w - \rho_o)g]|_{\eta=1}\right)$$
$$= \frac{M_2 k}{L}\left(\lim_{S_w \to 1-S_{nwr}} \lambda_w \frac{\partial S}{\partial \eta}\right). \tag{1.25}$$

The relative permeability for the wetting phase, k_{rw} vanishes in the limit $S_w = 1 - S_{nwr}$. In order to keep $u > 0$, $\partial S/\partial \eta$ has to become infinite. This indicates *boundary layer* behavior at the outlet boundary.

As initial conditions we take:

Imbibition:
$$S(\eta)|_{\tau=0} = S_{wc}. \tag{1.26}$$

Drainage:
$$S(\eta)|_{\tau=0} = 1 - S_{nwr}. \tag{1.27}$$

2 NUMERICAL SOLUTION. In order to predict water break through we need numerical schemes that keep numerical dispersion reasonably low. Unphysical oscillations are also unwanted. We illustrate this fact by showing the saturation profiles resulting from use of *the Galerkin method* with linear basis and test functions. It is well known that this scheme produces spurious oscillations, confer [2]. In Figure 2.1 a simulation of a convection-dominated problem is presented. The numerical solution oscillates around the sharp front. Even though the oscillations are quite stable before the 'front' reaches the outlet, the solution becomes unstable as the front arrives. The Galerkin method also produces an overshoot at the inlet boundary. This indicates that a refinement of the grid might not help. Therefore numerical oscillations should be avoided, or at least, one should be able to control them.

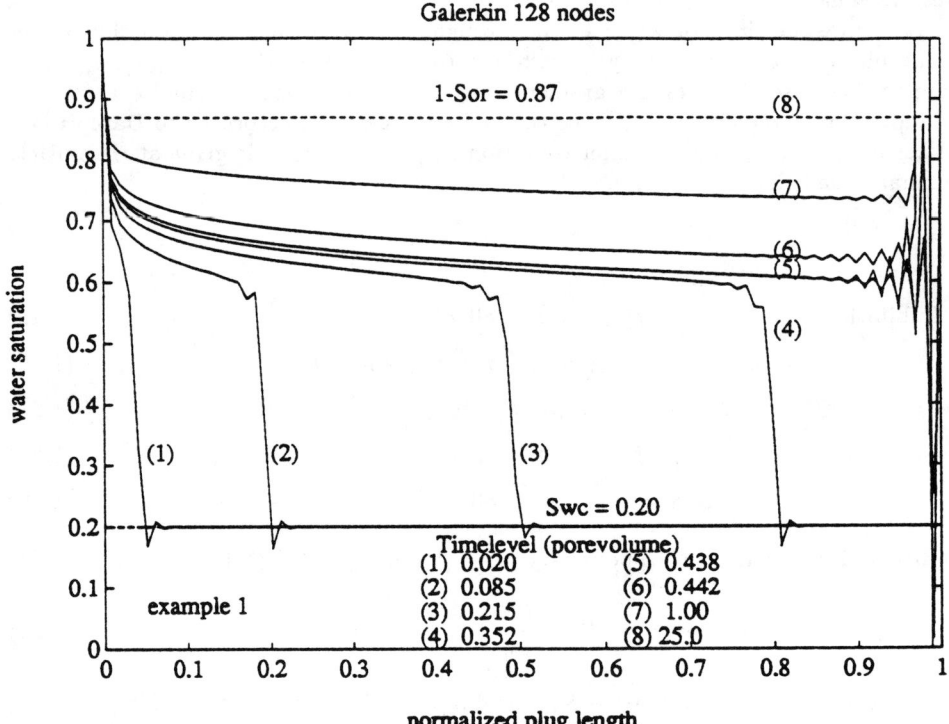

Figure 2.1 Galerkin with unphysical oscillations.

Figure 3.1 (in part 3) shows the *Petrov-Galerkin method* with linear basis and quadratic, asymmetric test functions. The oscillations have been removed at the expense of introducing *artificial diffusion*. In order to approach steady state it is necessary to take long time steps. This could not be achieved by explicit schemes, so for $\tau > \tau_b$ one should use an implicit scheme [3]. It is also observed that, for this problem, the convergence rate of the nonlinear iterations depends strongly upon the mass conservation. Therefore the discretization should satisfy material balance. In the following we present and compare two numerical approaches for solving the saturation equation (1.12) together with the conditions (1.19)-(1.21) and (1.26) or (1.22)-(1.23) and (1.27) numerically. The first approach is a time discretized Petrov-Galerkin method with linear basis functions and quadratic, asymmetric test functions. The scheme is implicit, and the nonlinear system arising at each time level is solved by the Newton-Raphson method [4]. The second is a *pseudospectral scheme* based on the ideas of *cell averages*, [5] [6]. We were guided to such an approach by the possibilities for spectral schemes to resolve sharp fronts, because an accurate description of those are needed in order to predict water break through. Also *the fast fourier transforms* involved will *vectorize* efficiently on a *vector machine*.

A Petrov-Galerkin method. The interval $[0, 1]$ (the normalized plug length) is divided into M elements. Element number e, noted Ω_e is defined by the closed interval $\eta \in [\eta_e, \eta_{e+1}]$, $e = 1, 2, ..., M$. To simplify further calculations, we transform element e to the unit element by:

$$\xi_e = \frac{\eta - \eta_e}{\eta_{e+1} - \eta_e}. \tag{2.1}$$

Let $\mathbf{H}^1(\Omega_e)$ denote the Sobolev space on Ω_e where the functions and their first order derivatives are square integrable.

On each Ω_e we have the following variational problem:

Find $S \in \mathbf{H}^1(\Omega_e)$ such that $\forall v \in \mathbf{H}^1(\Omega_e)$:

$$h_e \int_{\Omega_e} \phi \frac{\partial S}{\partial \tau} \, v \, d\xi_e = -fv|_0^1 + \int_{\Omega_e} f_c(S) \, u \frac{\partial v}{\partial \xi_e} d\xi_e$$

$$+ \frac{1}{h_e} \int_{\Omega_e} k \frac{\partial f_d}{\partial \xi_e} \frac{\partial v}{\partial \xi_e} d\xi_e, \tag{2.2}$$

$$h_e = \eta_{e+1} - \eta_e. \tag{2.3}$$

We now discretize the variational formulation on each element by approximating the solution in the finite-dimensional subspace $M^1(\Omega_e) \subset \mathbf{H}^1(\Omega_e)$ consisting of piece wise linears.

On each element e the saturation is interpolated by:

$$S_h^e = S_1^e v_1^e(\xi) + S_2^e v_2^e(\xi), \tag{2.4}$$

with basis:
$$v_1^e(\xi) = (1 - \xi), \tag{2.5}$$
$$v_2^e(\xi) = \xi. \tag{2.6}$$

For the test space we choose the asymmetric basis

$$T_1^e(\xi) = v_1^e(\xi) + 3\alpha^e \xi(1 - \xi), \tag{2.7}$$
$$T_2^e(\xi) = v_2^e(\xi) - 3\alpha^e \xi(1 - \xi). \tag{2.8}$$

where α^e is chosen to introduce artificial diffusion in order to avoid unphysical oscillations, [2] [7].

On each element the interpolant S_h^e is substituted for S and the saturation equation is multiplied by the test functions (2.7)-(2.8) and integrated over each element. The flux term is integrated by parts and the two-point trapezoidal rule is used for the numerical integration of the flux term. This gives the following element equations for element e:

$$f_1^e = \frac{h^e}{s} \left(\frac{1}{3} - \frac{\alpha^e}{4}\right) \frac{dS_1^e}{dt} + \frac{h^e}{s} \left(\frac{1}{6} - \frac{\alpha^e}{4}\right) \frac{dS_2^e}{dt}$$

$$+ \frac{1 - 3\alpha^e}{2} f_c(S_1^e) + \frac{1 + 3\alpha^e}{2} f_c(S_2^e)$$

$$+ \frac{1 - 3\alpha^e}{2} \Psi(S)|_{S_1^e}(S_2^e - S_1^e) \frac{1}{h^e}$$

$$+ \frac{1 + 3\alpha^e}{2} \Psi(S)|_{S_2^e}(S_2^e - S_1^e) \frac{1}{h^e} - f|_{\xi=0}, \tag{2.9}$$

$$f_2^e = \frac{h^e}{s} \left(\frac{1}{6} + \frac{\alpha^e}{4}\right) \frac{dS_1^e}{dt} + \frac{h^e}{s} \left(\frac{1}{3} + \frac{\alpha^e}{4}\right) \frac{dS_2^e}{dt}$$

$$- \frac{1 - 3\alpha^e}{2} f_c(S_1^e) - \frac{1 + 3\alpha^e}{2} f_c(S_2^e)$$

$$- \frac{1 - 3\alpha^e}{2} \Psi(S)|_{S_1^e}(S_2^e - S_1^e) \frac{1}{h^e}$$

$$- \frac{1 + 3\alpha^e}{2} \Psi(S)|_{S_2^e}(S_2^e - S_1^e) \frac{1}{h^e} + f|_{\xi=1}, \tag{2.10}$$

$$\Psi(S) = \frac{kH(S)}{L} \frac{dP_c}{dS}. \tag{2.11}$$

We have used Euler backwards for time discretization: (S_j° is the solution at the previous time level.)

$$\frac{dS_j}{d\tau} \sim \frac{S_j - S_j^\circ}{\triangle \tau} \tag{2.12}$$

The global system is defined through the coupling process:

$$\begin{aligned}
F_1 &= f_1^1, \\
F_j &= f_2^{j-1} + f_1^j, \qquad j = 2, 3, ..., M, \\
F_{M+1} &= f_2^M.
\end{aligned} \tag{2.13}$$

An important property of the test functions (2.7)-(2.8) is that they guarantee element wise *mass conservation*. Global mass conservation is guaranteed by the coupling procedure (2.13) since the boundary integrals cancel off, $f|_{\xi_2^{s-1}} = f|_{\xi_1^s}$

A pseudo-spectral scheme for the saturation equation. The scheme will be built on the ideas of *cell averages*. Integrate the saturation equation (1.12) over the interval $[\eta_{j-\frac{1}{2}}, \eta_{j+\frac{1}{2}}]$:

$$\int_{\eta_{j-\frac{1}{2}}}^{\eta_{j+\frac{1}{2}}} \frac{\partial S(\eta)}{\partial \tau} d\eta + (f_{j+\frac{1}{2}} - f_{j-\frac{1}{2}}) = 0. \tag{2.14}$$

$$\text{Define}: \bar{S}_j = \frac{1}{h} \int_{\eta_{j-\frac{1}{2}}}^{\eta_{j+\frac{1}{2}}} S(\eta) d\eta \quad \text{(cell average)}, \tag{2.15}$$

$$h = \eta_{j+\frac{1}{2}} - \eta_{j-\frac{1}{2}}. \tag{2.16}$$

Eq. (2.14) may be written as:

$$\frac{\partial \bar{S}_j}{\partial \tau} + \frac{1}{h}(f_{j+\frac{1}{2}} - f_{j-\frac{1}{2}}) = 0. \tag{2.17}$$

A major problem with cell average schemes is that we need point values for the flux f. This flux is again a function of S and its spatial derivative. At each time step we get a *reconstruction problem* for the point values. This is a serious problem, especially in several spatial dimensions. For spectral methods this is more straight forward, because the reconstruction is a convolution. We use the fast Fourier transform (**FFT**) to obtain the point values of S and its spatial derivative.

Let $\zeta_j = \frac{i-1}{N}$, $j = 1, 2, ..., N$ be the Fourier collocation points for $\zeta \in [0, 1]$.

Let $\qquad \bar{a}_k = \mathbf{FFT}(\bar{S})$ denote the transform

$$\bar{a}_k = \frac{1}{N} \sum_{j=0}^{N-1} \bar{S}(\zeta_j) e^{-2\pi i k \zeta_j}, \quad -\frac{N}{2} \le k \le \frac{N}{2} - 1. \tag{2.18}$$

Let $\qquad \bar{S}_j = \mathbf{FFT}^{-1}(\bar{a}_k)$ denote the inverse transform:

$$\bar{S}_j = \sum_{k=-\frac{N}{2}}^{\frac{N}{2}-1} \bar{a}_k e^{2\pi i k \zeta_j} \quad j = 1, 2, ..., N - 1. \tag{2.19}$$

The *Fourier collocation derivative* [8] is defined by:

$$\frac{\partial S}{\partial \zeta}\Big|_{\zeta_j} = \sum_{k=-\frac{N}{2}}^{\frac{N}{2}-1} i k a_k e^{2\pi i k \zeta_j}. \tag{2.20}$$

It is straight forward to derive that the cell averages and the point values are related as follows:

$$\bar{S}_j = \sum_{k=-\frac{N}{2}}^{\frac{N}{2}-1} \bar{a}_k e^{2\pi i k \zeta_j} \quad j = 1, 2, \cdots, N-1, \tag{2.21}$$

$$S_{j+\frac{1}{2}} = \sum_{k=-\frac{N}{2}}^{\frac{N}{2}-1} a_k e^{2\pi i k \zeta_{j+\frac{1}{2}}} \quad j = 1, 2, \cdots, N-1, \tag{2.21}$$

$$\bar{a}_k = \sigma_k a_k,$$

$$\sigma_k = \sin\frac{(\pi k h)}{\pi k h}, \quad 0 < |k| \leq N/2, \tag{2.23}$$

$$\sigma_0 = 1.$$

The above indicates that the Fourier coefficients a_k associated with the point values could be written using the Fourier coefficients for the cell averages. One is led to the following procedure for the reconstruction problem:

from $\bar{S}_j,$ $j = 1, 2, ..., N-1,$

get: $\bar{a}_k$ $= \mathbf{FFT}(\bar{S}_j) \quad -\frac{N}{2} \leq k \leq \frac{N}{2} - 1,$ $\tag{2.24}$

form: a_k $-\frac{N}{2} \leq k \leq \frac{N}{2} - 1 \quad$ (by using (2.22)),

get: $S_{j+\frac{1}{2}}$ $= \mathbf{FFT}^{-1}(a_k),$ $\tag{2.25}$

       $\frac{\partial S}{\partial \zeta}\big|_{j+\frac{1}{2}}$ $= \mathbf{FFT}^{-1}(ik a_k).$ $\tag{2.26}$

Time discretization. We have used the following *Runge-Kutta scheme* [9] for forward time-stepping . Let the solution be known at time level ν; $\bar{S}_j|_\nu = \bar{S}_j^\nu$. Then we proceed to the next time level $\nu + 1$ by:

Let

$$\bar{Z}_j \quad = \bar{S}_j^\nu, \quad j = 1, 2, \cdots, N-1. \tag{2.27}$$

For $l = m, 1, -1$:

$$\bar{Z}_j \quad = \bar{S}_j^\nu + \frac{1}{l}(\tau_{v+1} - \tau_v)(f(Z_{j+\frac{1}{2}}) - f(Z_{j-\frac{1}{2}})),$$
$$j = 1, 2, \cdots, N-1. \tag{2.28}$$

Loop end.

Set

$$\bar{S}_j^\nu \quad = \bar{Z}_j, \quad j = 1, 2, \cdots, N-1. \tag{2.29}$$

Eqs. (2.27)-(2.29) define a Runge-Kutta method of order m. In this work we have used $m = 6$.

Treatment of non periodic boundary conditions. The Fourier method is valid only for periodic problems. Clearly, by conditions (1.19)-(1.23), we are dealing with a non periodic problem. We use the following procedure: At each side of the boundaries define an *artificial extension* of the physical domain. The computational domain is now defined by:

$$\zeta \in [0, a > \cup [a, b] \cup < b, 1], \quad 0 < a < b < 1. \tag{2.30}$$

The interval $[a, b]$ is the physical interval and it is transformed into the computational domain:

$$\zeta = \eta \cdot (a - b) + b, \quad \eta \in [0, 1]. \tag{2.31}$$

The artificial extensions will act as 'buffers' for *the Gibbs phenomena* which is to be expected at the boundaries because of non periodicity. In our work, we found experimentally that $a = 1 - b = 0.1$ was a choice that enabled us to deal with non periodicity.

A slope limiter. For convection-dominated flows the scheme described by (2.17) together with the reconstruction procedure (2.24)-(2.26) and the Runge-Kutta time stepping procedure (2.27)-(2.29) may give unphysical oscillations in the neighborhood of sharp fronts. To stabilize the scheme we use a *slope limiter* introduced by Van Leer [10] and modified by Chavent and Jaffre [11].

For each of the intervals $[\zeta_i, \zeta_{i+1}]$, introduce the average values

$$S_i^* = \frac{1}{2}(\bar{S}_i + \bar{S}_{i+1}), \quad i = 1, 2 ..., N - 1, \quad (\bar{S}_N = \bar{S}_1). \tag{2.32}$$

In order to modify $\bar{S}_i$, solve the following minimization problems for $i = 1, 2, \cdots, N-1$:

Find V_i^+ and V_{i+1}^- that minimize

$$(V_i^+ - \bar{S}_i)^2 + (V_{i+1}^- - \bar{S}_{i+1})^2$$

subject to

$$(1 - \theta)S_i^* + \theta \min(S_{i-1}^*, S_i^*) \le V_i^+ \le (1 - \theta)S_i^* \tag{2.33}$$
$$+ \theta \max(S_{i-1}^*, S_i^*), \tag{2.34}$$

$$(1 - \theta)S_i^* + \theta \min(S_i^*, S_{i+1}^*) \le V_{i+1}^- \le (1 - \theta)S_i^* \tag{}$$
$$+ \theta \max(S_i^*, S_{i+1}^*), \tag{2.35}$$

$$S_i^* = \frac{1}{2}(V_i^+ + V_{i+1}^-). \tag{2.36}$$

Modify the solution $\bar{S}_i$ according to:

$$\bar{S}_i = \frac{1}{2}(V_i^- + V_i^+). \tag{2.37}$$

(2.36) and (2.37) guarantees mass conservation. By increasing θ from zero to 1 we obtain a family of schemes, with decreasing artificial diffusion.

We found experimentally that a good choice for θ was $\theta \approx 3/5$. For all our numerical simulation tests this gave a non oscillatory solution with sharp fronts. (Chavent and Jaffre found $\theta \approx 2/5$ as the preferred choice for the *discontinuous finite element scheme* described in [11].)

Let us now recapitulate the procedure needed to obtain the solution at time level $\nu + 1$, noted $\{\bar{S}_j^{\nu+1}\}$, from the solution at time level ν, $\{\bar{S}_j^\nu\}$:

(1) Solve the reconstruction problem described by (2.24)-(2.26) in order to find the point values $\{S_{j+\frac{1}{2}}^\nu\}$ and $\{\partial S_{j+\frac{1}{2}}^\nu / \partial \zeta\}$.

(2) Use the Runge-Kutta scheme described by (2.27)-(2.29) to find $\{\bar{S}_j^{\nu+1}\}$

(3) Apply the slope limiter described by (2.33)-(2.37) to modify $\{\bar{S}_j^{\nu+1}\}$

(4) Put $\bar{S}_j = \bar{S}_{j_a} = \bar{S}_a \ \forall \ \zeta_j \in [0, a >$ and $\bar{S}_j = \bar{S}_{j_b} = \bar{S}_b \ \forall \ \zeta_j \in < b, 1]$

3 TEST EXAMPLES. We have used the following expressions for defining the relative permeabilities and the capillary pressure as functions of the water saturation, S_w:

$$k_{rw} = \frac{k_w}{k} = \frac{k'_w}{k} \Big(\frac{S_w - S_{wc}}{1 - S_{or} - S_{wc}}\Big)^{d_w}, \qquad (3.1)$$

$$k_{ro} = \frac{k_o}{k} = \frac{k'_o}{k} \Big(\frac{1 - S_{or} - S_w}{1 - S_{or} - S_{wc}}\Big)^{d_o}, \qquad (3.2)$$

$$P_c = \bar{P}_c \Big(\frac{1 - S_{or} - S_w}{S_w - S_{wc}}\Big)^{d_c}. \qquad (3.3)$$

k_o and k_w are the *effective permeabilities* for oil and water.

k'_o and k'_w are the *effective end point permeabilities* for oil and water.

d_o and d_w are the *relative permeability exponents* for oil and water.

$\bar{P}_c$ is the *capillary pressure coefficient*.

d_c is the *capillary pressure exponent*.

DATA common to test ex.no.1 and 2:		
Variable	Value	SI-units
S_{wc}	0.20	
S_{init}	0.20	
S_{or}	0.13	
ϕ	0.173	
L	63.7	millimeters (mm)
A	1130.0	millimeters squared (mm^2)
ρ_w	1000.0	kilogram/cubic meters (kg/m^3)
ρ_o	800.0	kilogram/cubic meters (kg/m^3)
μ_w	1.00	millipascal/second (10^{-3}Pa)
μ_o	2.24	millipascal/second (10^{-3}Pa)
k'_w	0.20	micrometers squared (10^{-12}m^2)
k'_o	0.52	micrometers squared (10^{-12}m^2)
$\bar{P}_c$	3.9	kilopascal (10^3 Pa)
d_c	0.4749	
g_x	0.0	

Table 3.1 Data common to test example 1 and 2.

We are going to concentrate on imbibition experiments. The relative error in material balance (mass of water injected minus mass of water accumulated divided by mass of water injected) will be less than $0.5 \cdot 10^{-6}$ for the cases shown. (Checked before mobile water reaches the outlet.)

We have considered one convection-dominated problem for which the saturation profiles exhibit steep fronts close to a shock solution, and one problem for which the profiles are smeared out by the capillary pressure. Data for these two cases are given in Table 3.1 and Table 3.2.

For the Petrov-Galerkin (PG) method we need $2N$ function evaluations to build the Jacobian matrix at each Newton-Raphson iteration since we calculate the Jacobian matrix by one sided differences. For the solution of the linear system we use *Thomas algorithm* which takes $6N$ multiplications and divisions. Let the number of Newton-Raphson iterations pr.time step be n. For our simulations, n is between 3 and 7.

For the Pseudospectral (PS) method we need one Fourier transform and two inverse transforms for the reconstruction problem defined by (2.23)-(2.25). Each transform is of $O(N \log N)$. The Runge-Kutta scheme (2.26)- (2.28) of order m uses $m(N-1)$ function evaluations. In addition, some extra work is done by the slope limiter.

Volumetric flow rate and relative permeability exponents			
Volumetric flow rate ($\mathcal{V} = u\mathcal{A}$)		d_w	d_o
Test ex.no.1:	100.0 cubic millimeter/second	4	5
Test ex.no.2:	10.0 cubic millimeter/second	2	3

Table 3.2 Volumetric flow rate and relative permeability exponents

for test examples 1 and 2.

Result of simulations. In Table 3.3 we have shown the CPU-time for one Newton-Raphson iteration of the Petrov-Galerkin method compared to the CPU-time needed for one time step of the pseudospectral method. These runs were done with the data for test example 1.

IBM-3090 VF uniprocessor CPU (seconds):		
Nodes	PG	PS
64	0.070s	0.025s
128	0.112s	0.046s
256	0.201s	0.087s
512	0.415s	0.167s

Table 3.3 CPU times for one Newton-Raphson iteration (PG)

and one time step (PS)

The pseudospectral scheme is about $2.5n$ times faster than the Petrov-Galerkin method pr.timestep. (Compiler: IBM VS-FORTRAN 2.3 vector level 2, vector length 128.) With the implicit Petrov-Galerkin scheme one may take longer time steps, the only practical restriction is the convergence of the Newton-Raphson iterations. On the other hand, the pseudospectral scheme gives higher spatial accuracy.

Explanation

NO	Number of nodes.
$\triangle S$	If specified, maximum allowed saturation change pr. time step. This value represents 'the stability limit' for the pseudospectral scheme: $2 \triangle S$ results in unstable behavior.
PG	Petrov-Galerkin's method, $\alpha = -\frac{2}{3}$.
PS	Pseudospectral method with sixth order Runge-Kutta.
τ_a	Time for arrival at the outlet.
τ_b	Time for water break through.
CPU	CPU-time, IBM 3090 VF uniprocessor. (xx:yy denotes xx minutes and yy seconds.)

N	$\triangle S$	method	τ_a	τ_b	CPU $\tau = \tau_b$	CPU $\tau = 25.0$
1024	$0.10,\ (\tau_a \leq \tau \leq \tau_b)$	PG	.433	.436	$116:07$	
512	$0.10,\ (\tau_a \leq \tau \leq \tau_b)$	PG	.431	.434	$28:27$	$36:32$
256	$0.10,\ (\tau_a \leq \tau \leq \tau_b)$	PG	.426	.432	$07:08$	$09:08$
128	$0.10,\ (\tau_a \leq \tau \leq \tau_b)$	PG	.419	.428	$01:34$	$03:09$
64	$0.10,\ (\tau_a \leq \tau \leq \tau_b)$	PG	.408	.424	$00:22$	$01:04$
512	0.10	PS	.434		$07:16$	
256	0.10	PS	.433		$02:08$	
128	0.10	PS	.431		$00:23$	
64	0.10	PS	.417		$00:06$	

Table 3.4 Results for test ex.no.1

NO	$\triangle S$	method	τ_a	τ_b	CPU $\tau = \tau_b$	CPU $\tau = 25.0$
1024		PG	.254	.392	$49:07$	
512		PG	.251	.393	$10:01$	$16:08$
256		PG	.245	.393	$02:14$	$07:12$
128		PG	.243	.396	$00:24$	$01:29$
64		PG	.235	.400	$00:12$	$00:25$
512	0.002	PS	.254		$149:22$	
256	0.002	PS	.254		$17:20$	
128	0.002	PS	.250		$02:15$	
64	0.002	PS	.242		$00:22$	

Table 3.5 Results for test ex.no.2

For both test examples, the Petrov-Galerkin (PS) with 2048 nodes was taken as the reference solution. (4096 nodes gave identically times for τ_a and τ_b.) It is evident from Table 3.3 and Table 3.4 that the pseudospectral scheme gives a better prediction of the time τ_a for arrival of mobile water at the outlet than the Petrov-Galerkin scheme. For test example 1 the pseudospectral scheme is about 4 times faster than the Petrov-Galerkin scheme. Figure 3.5 and Figure 3.6 show the positioning of the saturation profiles at $\tau = 0.20$ for test example 1. The pseudospectral scheme models the position and sharpness of the front very accurate (spectral accuracy). The Petrov-Galerkin scheme gives more smearing, a smaller absolute value of α may lead to oscillatory behavior [12]. For test example 2 the Petrov-Galerkin scheme 'wins' with respect to CPU time. This is due to the restrictive maximum saturation change pr. time step ($\triangle S$) for the pseudospectral scheme in this case. The small time step allowed is caused by the second-order elliptic term in the differential equation. (For explicit difference schemes the diffusive term causes a time step restriction of $O(h^2)$ while the convective term gives a restriction of $O(h)$.)

CONCLUSION. We have demonstrated that the Petrov-Galerkin method and the pseudospectral scheme described in this work may be used in the modelling of laboratory displacement experiments. Both schemes conserve mass exactly. With the implicit Petrov-Galerkin method one could take long time steps to approach steady state.

The explicit pseudospectral scheme may be used before water reaches the outlet. The front velocity and sharpness are more accurate compared to the Petrov-Galerkin method. With reference to the test examples presented in this work, it may be used with advantage for convection-dominated flows. Problems with a larger amount of diffusion give a restrictive condition on the time step.

REFERENCES

1. C. M. MARLE, *Multiphase flow in porous media*, Éditions Technip, Paris, 1981.

2. A. R. MITCHELL and D. T. GRIFFITHS *The Finite Difference Method in Partial Differential Equations*, John Wiley & Sons, New York, 1980.

3. I. AAVATSMARK, *Modelling of laboratory displacement experiments*, Z. angew. Math. Mech. 71 (1991) 9, pp. 341-354. (In German.)

4. K. AZIZ and A. SETTARI, *Petroleum reservoir simulation*, Applied Science Publishers Ltd., London, 1979.

5. A. HARTEN, B. ENGQUIST, S. OSHER and S. R. CHAKRAVARTHY (1978), *Uniformly High Order Accurate Essentially Nonoscillatory Schemes, iii,*, J. o. Comp. Phys. Volume 71, Number 2 (August 1978), pp. 231-303.

6. W. CAI, D. GOTTLIEB and C. W. SHU, *Non-oscillatory spectral fourier methods for shock wave calculations*, ICASE REPORT NO. 88-37, 1988.

7. J. W. BARRET and K. W. MORTON, *Approximate symmetrization and Petrov-Galerkin methods for diffusion-convection problems*, Comput. Meths. Appl. Mech. Engrg., 45 (1984), pp. 97-122.

8. C. CANUTO, M. Y. HUSSANI, A. QUARTERONI and T. A. ZANG, *Spectral Methods in Fluid Dynamics*, Springer Verlag, New York, 1988.

9. A. JAMESON, H. SCHMIDT and E. TURKEL, *Numerical Solutions of the Euler Equations by Finite Volume Methods Using Runge-Kutta Time Stepping Schemes*, AIAA Pap. No. 81-1259, 1981.

10. B. VAN LEER, *Towards the ultimate conservative scheme : iv. A new approach to numerical convection*, J. Comp. Phys. 23 (1977), pp. 276-299.

11. G. CHAVENT and J. JAFFRE, *Mathematical Models and Finite Elements for Reservoir Simulation*, North Holland, 1987.

12. Ø. BØE, *Application Of a Time Discretized Petrov-Galerkin Method for Simulation of Laboratory Displacement Experiments*, Bergen Scientific Centre Report Series 87/15, 1987.

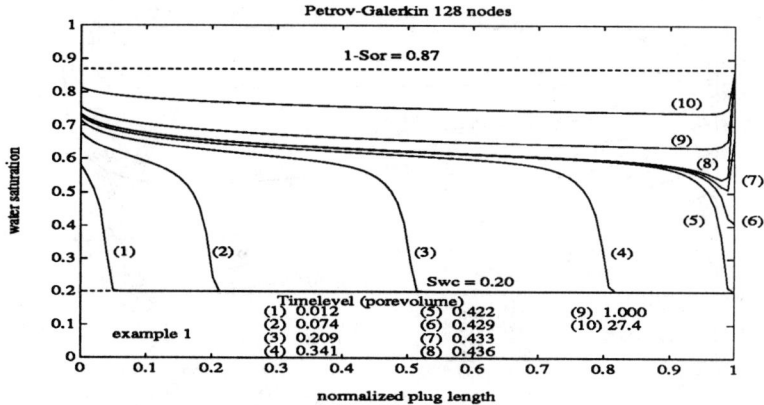

Figure 3.1 Petrov-Galerkin, test example 1.

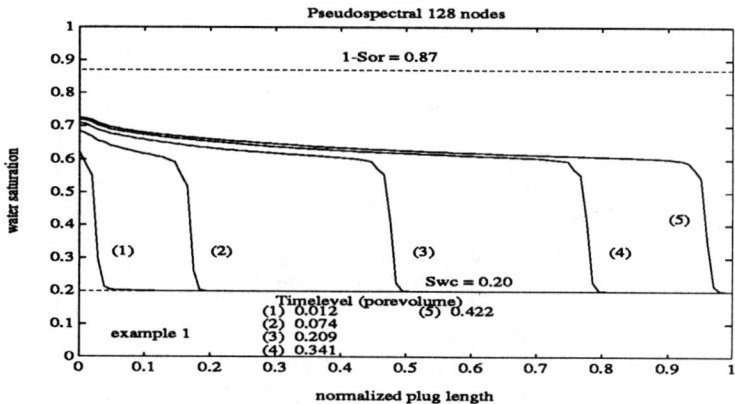

Figure 3.2 Pseudospectral, test example 1.

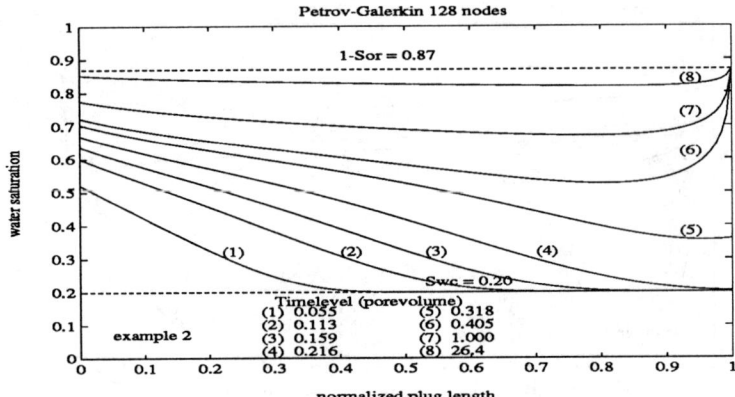

Figure 3.3 Petrov-Galerkin, test example 2.

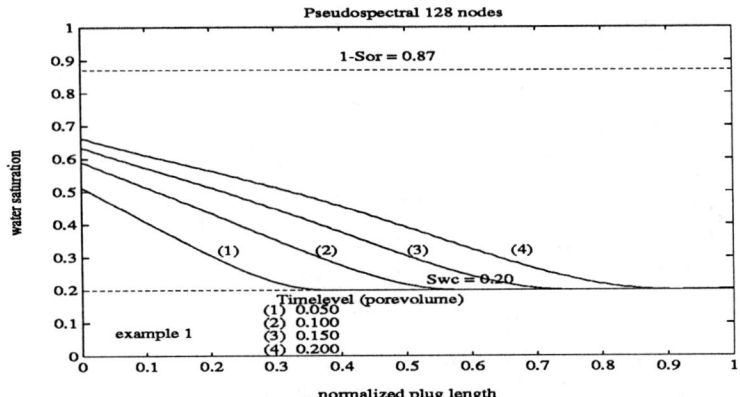

Figure 3.4 Pseudospectral, test example 2.

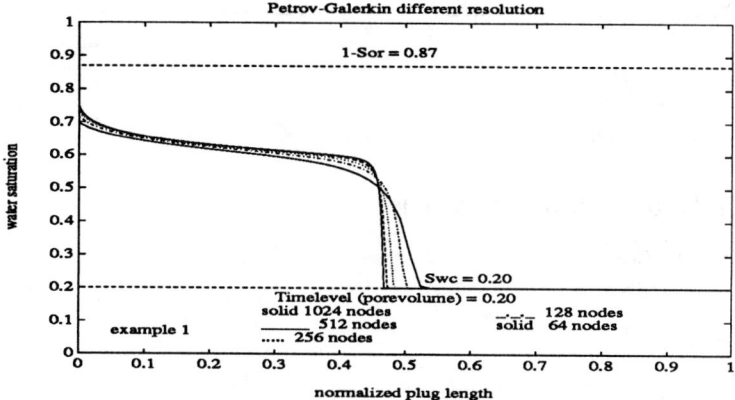

Figure 3.5 Petrov-Galerkin, different number of nodes, test example 1.

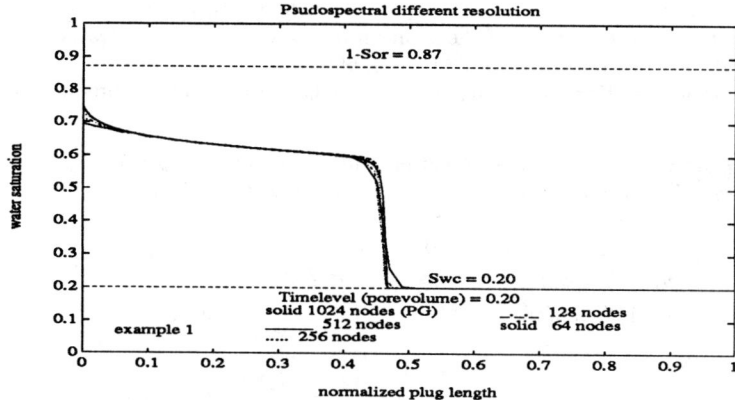

Figure 3.6 Pseudospectral, different number of nodes, test example 1.

CHAPTER 3

Semi-Linear Heat Equation Subject to the Specification of Energy*

John R. Cannon**
Yanping Lin†
John Van der Hoek‡

Abstract. In this paper we shall study the following problem: $u_t = u_{xx} + F(x, t, u, u_x)$, $0 < x < 1$, $0 < t \leq T$; $u(x, 0) = \phi(x)$, $0 < x < 1$; $u_x(1, t) = g(t)$ or $u(1, t) = g(t)$, $0 < t \leq T$, and $E(t) = \int_0^{s(t)} u(x, t) \, dx$, $0 < t \leq T$, $0 < s(t) \leq 1$. The existence, uniqueness and continuous dependence upon the data θ, g, E, F of the solution for fixed s are demonstrated.

AMS(MOS) subject classification: 65N15, 65N30.

Keywords: Heat equation specification of energy, Green function, Neumann function.

1. Introduction. In some heat conduction and other diffusion processes, the following non-classical boundary value problem is of interest. Find $u = u(x, t)$ such that

(1.1) $\qquad u_t = u_{xx} + F(x, t, u, u_x), \qquad 0 < x < 1, \qquad 0 < t \leq T;$

(1.2) $\qquad u(x, 0) = \phi(x), \qquad\qquad 0 < x < 1;$

(1.3) $\qquad u_x(1, t) = g(t), \qquad\qquad\qquad\qquad 0 < t \leq T;$

and

(1.4) $\qquad\qquad \int_0^{s(t)} u(x, t) \, dx = E(t), \quad 0 < t \leq T, \quad 0 < s(t) \leq 1,$

where ϕ, g, E, s and F are known functions. If we know that u represents the temperature of a slab or bar, then (1.1)-(1.4) is a simple model of heat conduction with heat energy prescribed on a portion of the physical body $0 \leq x \leq s(t)$, $0 < t \leq T$.

When $F \equiv 0$, problem (1.1)-(1.4) was first investigated by Cannon [1] in 1963. Kamyin [15] extended the result of [1] to the general linear parabolic equation in 1964. The problems similar to (1.1)-(1.4) with $F = F(x, t)$ have also been studied by several authors [3, 9, 10, 11, 13, 14, 19]. A free boundary value problem related to (1.4) has also been studied by Cannon and Van der Hoek in [5, 6] while numerical approximations (finite difference and finite element) to the solution are considered in [4, 7, 8]. Here we shall consider the non-linear case of $F = F(x, t, u, u_x)$ and show that problem (1.1)-(1.4) has a unique solution which is continuously dependent upon the data when $s(t)$ is fixed.

This paper is organized in the following way. In Section 2 we shall transform condition (1.4) to an equivalent form which is easy to handle and begin our study there. The existence and uniqueness

*This research was supported in part by the National Science Foundation Grant DMS-8901301 and Texas Higher Education Coordinating Board Grant 003581-006. This research was supported in part by the NSERC of Canada.
**Department of Mathematics, Lamar University, Beaumont, TX 77710.
†Department of Mathematics, University of Alberta, Edmonton, Canada, T6G 2G1.
‡Department of Applied Mathematics, University of Adelaide, Adelaide, South Australia, 5001, Ausytralia.

of the solution will be presented in Section 3 while continuous dependence is discussed in Section 4. In Section 5 we shall consider the first boundary value case:

$$(1.5) \qquad u_t = u_{xx} + F(x,t,u,u_x), \qquad 0 < x < 1, \qquad 0 < t \leq T;$$

$$(1.6) \qquad u(x,0) = \phi(x), \qquad 0 < x < 1;$$

$$(1.7) \qquad u(1,t) = g(t), \qquad 0 < t \leq T;$$

$$(1.8) \qquad \int_0^{s(t)} u(x,t)\,dx = E(t), \qquad 0 < s(t) < 1, \qquad 0 < t \leq T.$$

Problem (1.5)-(1.8) is similar to problem (1.1)-(1.4), but requires some extra work.

Assumption H1:

$$\phi \in C^1[0,1], \quad \int_0^{s(0)} \phi(x)\,dx = E(0), \quad s,g,E \in C^1([0,T]),$$

$$\phi'(1) = g(0), \quad \phi'(0) = \phi(x(0))s'(0) + \phi'(s(0)) - E'(0) + \int_0^{s(0)} F(x,0,\phi,\phi')\,dx$$

$$F \in C(\bar{Q}_T \times R^2) \text{ and there exists } L > 0 \text{ such that}$$

$$\bigl|F(x,t,p,q) - F(x,t,p',q')\bigr| \leq L\bigl(|p - p'| + |q - q'|\bigr).$$

Definition 1.1: A function $u(x,t)$ is called a solution of problem (1.1)-(1.4) or problem (1.5)-(1.8) if

    i): u, u_x are continuous in $0 \leq x \leq 1$, $0 \leq t \leq T$;

    ii): u_t, u_{xx} are continuous in $0 < x < 1$, $0 < t \leq T$;

    iii): (1.1)-(1.4) or (1.5)-(1.8) is satisfied.

The following notation will be used throughout this paper. For any function $u(x,t)$ defined on $0 < x < 1$, $0 < t \leq T$, we set

$$\|u\|_\infty = \operatorname*{ess\,sup}_{0<x<1,\ 0<t\leq T} |u(x,t)|,$$

$$\|u(\cdot,t)\| = \operatorname*{ess\,sup}_{0<x<1} |u(x,t)|,$$

$$\|u\|_1 = \|u\|_\infty + \|u_x\|_\infty.$$

It is easy to see that

$$\|u\|_\infty = \operatorname*{ess\,sup}_{0<t\leq T} \|u(\cdot,t)\|.$$

We also define

$$|f|_t = \sup_{0<r\leq t} |f(r)|$$

for any function defined on $0 < t \leq T$. For $\phi \in C^1([0,1])$ and $g \in C^1([0,T])$ we also use $\|\cdot\|_1$ to denote

$$\|\phi\|_1 = \|\phi\|_\infty + \|\phi'\|_\infty \text{ and } \|g\|_1 = \|g\|_\infty + \|g'\|_\infty.$$

2. An equivalent problem. Formally, let us differentiate (1.4) with respect to t to obtain

$$(2.1) \qquad u_x(0,t) = u\bigl(s(t),t\bigr)s'(t) + u_x\bigl(s(t),t\bigr) + \int_0^{s(t)} F\bigl(x,t,u(x,t),u_x(x,t)\bigr)\,dx - E'(t).$$

Definition 2.1: A function $u = u(x,t)$ is called a solution of (1.1)-(1.3) and (2.1) if it satisfies (i)-(ii) of definition 1.1 and (1.1)-(1.3) and (2.1).

Now we give a lemma which implies that (1.4) is equivalent to (2.1).

Lemma 2.1. Under assumption H1, the problem (1.1)-(1.4) is equivalent to the problem (1.1)-(1.3) and (2.1).

Proof: If u is a solution of (1.1)-(1.4), then it is also a solution of (1.1)-(1.3) and (2.1) by the derivation of (2.1). On the other hand, suppose that u is a solution of (1.1)-(1.3) and (2.1), let $\epsilon, \sigma > 0$ be small and $D_{\epsilon,\sigma} = \{(x,\tau) \mid 0 < \epsilon \leq x \leq s(t),\ 0 < \sigma \leq \tau \leq t\}$. We integrate (1.1) on $D_{\epsilon,\sigma}$ to obtain

$$(2.2) \qquad \int_{D_{\epsilon,\sigma}} u_t \, dx \, d\tau = \int_{D_{\epsilon,\sigma}} u_{xx} \, dx \, d\tau + \int_{D_{\epsilon,\sigma}} F(x,t,u,u_x) \, dx \, d\tau.$$

Since

$$(2.3) \qquad \int_{D_{\epsilon,\sigma}} u_t \, dx \, d\tau = \int_{\epsilon}^{s(t)} u(x,t) \, dx - \int_{\epsilon}^{s(\sigma)} u(x,\sigma) \, dx - \int_{\sigma}^{t} s'(\tau) u(s(\tau),\tau) \, d\tau$$

and

$$(2.4) \qquad \int_{D_{\epsilon,\sigma}} u_{xx} \, dx \, d\tau = \int_{\sigma}^{t} u_x(s(\tau),\tau) \, d\tau - \int_{\sigma}^{t} u_x(\epsilon,\tau) \, d\tau.$$

Substituting (2.3)-(2.4) into (2.2) and letting $\epsilon \to 0$, it follows that

$$(2.5) \qquad \int_{0}^{s(t)} u(x,t) \, dx - \int_{0}^{s(\sigma)} u(x,\sigma) \, dx - \int_{\sigma}^{t} s'(\tau) u(s(\tau),\tau) \, dx$$
$$= \int_{\sigma}^{t} u_x(s(\tau),\tau) \, d\tau - \int_{\sigma}^{t} u_x(0,\tau) \, dx + \int_{\sigma}^{t} \int_{0}^{s(\tau)} F(x,\tau,u(x,\tau),u_x(x,\tau)) \, dx \, d\tau.$$

It is easy to see by letting $\sigma \to 0$ in (2.5) using (2.1) that

$$(2.6) \qquad \int_{0}^{s(t)} u(x,t) \, dx - \int_{0}^{s(0)} u(x,0) \, dx = \int_{0}^{t} E'(\tau) \, d\tau = E(t) - E(0).$$

Thus, (1.4) follows from the compatibility assumption H1.   Q.E.D.

Now we consider a model problem

$$(2.7) \qquad \begin{array}{ll} u_t = u_{xx} + F(x,t,u,u_x), & 0 < x < 1,\ \ 0 < t \leq T; \\ u(x,0) = 0, & 0 < x < 1; \\ u_x(1,t) = 0, & 0 < t \leq T; \end{array}$$

and

$$(2.8) \qquad \int_{0}^{s(t)} u(x,t) \, dx = E(t),\ 0 < s(t) \leq 1,\ 0 < t \leq T,$$

where E, s and F satisfy assumption H1.

Lemma 2.2. If there exists a positive constant $\epsilon = \epsilon(T, L) > 0$ such that (2.7)-(2.8) has a local solution in $0 \leq t \leq \epsilon$, then problem (1.1)-(1.4) is solvable in $0 \leq t \leq T$ for any $T > 0$.

Proof: Consider an auxiliary problem

$$(2.9) \qquad \begin{array}{ll} v_t = v_{xx}, & 0 < x < 1,\ \ 0 < t \leq T; \\ v(x,0) = \phi(x), & 0 < x < 1; \\ v_x(1,t) = g(t), & 0 < t \leq T; \\ v_x(0,t) = \phi'(0), & 0 < t \leq T. \end{array}$$

It is well-known that (2.9) has a unique solution v such that v, v_x are continuous in $0 \leq x \leq 1$, $0 \leq t \leq T$. Let $w = u - v$, then it follows from (1.1)-(1.4) and (2.9) that

$$(2.10) \qquad \begin{array}{ll} w_t = w_{xx} + F(x,t,v+w,v_x+w_x), & 0 < x < 1,\ \ 0 < t \leq T; \\ w(x,0) = 0, & 0 < x < 1; \\ w_x(1,t) = 0, & 0 < t \leq T; \\ \int_{0}^{s(t)} w(x,t) \, dx = E(t) - \int_{0}^{s(t)} v(x,t) \, dx \in C^1[0,T], & 0 < t \leq T. \end{array}$$

From the hypothesis of this lemma and H1, there exists a w in $0 \le t \le \epsilon$ which is a solution of (2.10). Consequently, $u = v + w$ will be a solution of (1.1)-(1.4) in $0 \le t \le \epsilon$. Now let $z(x,t) = u(x, t + \epsilon)$, then (1.1)-(1.4) becomes

(2.11)
$$
\begin{aligned}
&z_t = z_{xx} + F(x, t + \epsilon, z, z_x), && 0 < x < 1, && 0 < t \le T; \\
&z(x, 0) = u(x, \epsilon), && 0 < x < 1; \\
&z_x(1, t) = g(t + \epsilon), && && 0 < t \le T; \\
&\int_0^{s(t+\epsilon)} z(x, t)\, dx = E(t + \epsilon), && && 0 < t \le T.
\end{aligned}
$$

We see from the proof given above that there exists a z in $0 \le t \le \epsilon$ for (2.11). Hence, the solution of (1.1)-(1.4) exists in $0 \le t \le 2\epsilon$. Repeating this procedure finitely many times, it follows that the solution of (1.1)-(1.4) exists in $0 \le t \le T$ for any $T > 0$.

Q.E.D.

Remark: Lemma 2.1 and Lemma 2.2 imply that it is sufficient to show that there exists a $\epsilon = \epsilon(T, L)$ such that problem (2.7)-(2.8) has a solution in $0 \le t \le \epsilon$ in order to show the global existence of the solution of (1.1)-(1.4).

We shall in next section to show that problem (2.7)-(2.8) has a unique local solution.

3. Existence and uniqueness. In this section we shall prove that there exists a unique solution for the following model problem

(3.1)
$$
\begin{aligned}
&u_t = u_{xx} + F(x, t, u, u_x), && 0 < x < 1, && 0 < t \le T; \\
&u(x, 0) = 0, && 0 < x < 1; \\
&u_x(1, t) = 0, && && 0 < t \le T;
\end{aligned}
$$

and

(3.2) $\quad u_x(0, t) = u\big(s(t), t\big)s'(t) + u_x\big(s(t), t\big) + \int_0^{s(t)} F\big(x, t, u(x, t), u_x(x, t)\big)\, dx - E'(t), \quad 0 < t \le T.$

Let u be a solution of (3.1)-(3.2). Then it follows from Cannon [2] that u can be represented uniquely as

(3.3) $\quad u(x, t) = -2 \int_0^t \theta(x, t - \tau) u_x(0, \tau)\, d\tau + \int_0^t \int_0^1 N(x, \xi, t, \tau) F\big(\xi, \tau, u(\xi, \tau), u_\xi(\xi, \tau)\big)\, d\xi\, d\tau,$

where

$$
\theta(x, t) = \sum_{m=-\infty}^{\infty} K(x + 2m, t), \quad t > 0,
$$

$$
K(x, t) = \frac{1}{\sqrt{4\pi t}} \exp\left\{-\frac{x^2}{4t}\right\}, \quad t > \tau,
$$

and

$$
N(x, \xi, t, \tau) = \theta(x - \xi, t - \tau) + \theta(x + \xi, t - \tau), \quad t > \tau.
$$

Thus, we have

(3.4)
$$
\begin{aligned}
u\big(s(t), t\big) = &-2 \int_0^t \theta\big(s(t), t - \tau\big) u_x(0, \tau)\, d\tau \\
&+ \int_0^t \int_0^1 N\big(s(t), \xi, t, \tau\big) F\big(\xi, \tau, u(\xi, \tau), u_\xi(\xi, \tau)\big)\, d\xi\, d\tau,
\end{aligned}
$$

(3.5)
$$
\begin{aligned}
u_x(x, t) = &-2 \int_0^t \theta_x(x, t - \tau) u_x(0, \tau)\, d\tau \\
&+ \int_0^t \int_0^1 N_x(x, \xi, t, \tau) F\big(\xi, \tau, u(\xi, \tau), u_\xi(\xi, \tau)\big)\, d\xi\, d\tau,
\end{aligned}
$$

(3.6)
$$
\begin{aligned}
u_x\big(s(t), t\big) = &-2 \int_0^t \theta_x\big(s(t), t - \tau\big) u_x(0, \tau)\, d\tau \\
&+ \int_0^t \int_0^1 N_x\big(s(t), \xi, t, \tau\big) F\big(\xi, \tau, u(\xi, \tau), u_\xi(\xi, \tau)\big)\, d\xi\, d\tau.
\end{aligned}
$$

Hence, (3.2) can be rewritten by using (3.4)-(3.6) as

$$(3.7) \qquad u_x(0,t) = \int_0^t \Phi(t,\tau)u_x(0,\tau)\,d\tau + \int_0^t \int_0^1 \Psi(\xi,t,\tau)F\big(\xi,\tau,u(\xi,\tau),u_\xi(\xi,\tau)\big)\,d\xi\,d\tau$$
$$+ \int_0^{s(t)} F\big(x,t,u(x,t),u_x(x,t)\big)\,dx - E'(t)$$

where

$$\Phi(t,\tau) = -2\theta\big(s(t),t-\tau\big)s'(t) - 2\theta_x\big(s(t),t-\tau\big),$$
$$\Psi(\xi,t,\tau) = N\big(s(t),\xi,t,\tau\big)s'(t) + N_x\big(s(t),\xi,t,\tau\big),$$

and where we have from Cannon [2] that there exists a constant $C > 0$ such that

$$(3.8) \qquad |\Phi(t,\tau)| \le \frac{C}{\sqrt{t-\tau}}, \quad t > \tau.$$

Lemma 3.1. Given (3.1), the condition (3.2) is equivalent to the condition (3.7).

Proof: If u is a solution of (3.1)-(3.2), then (3.7) follows from its derivation. On the other hand, if u is a solution of (3.1) and (3.7), it can be represented by (3.3), and $u(s(t),t)$, $u_x(x,t)$ and $u_x(s(t),t)$ can be expressed by (3.4), (3.5) and (3.6), respectively. Thus, (3.2) follows by (3.4)-(3.7).
$$\text{Q.E.D.}$$

Theorem 3.1. If s, E and F satisfy assumption H1, then there exists an $\epsilon = \epsilon(T,L)$ such that the problem (3.1) and (3.7) has a unique solution in $0 \le t \le \epsilon$.

Proof: We shall employ the following iteration procedure. Let $u^0 = 0$ and $\{u^n\}$ be defined as follows: If u^{n-1} is known, then solve

$$(3.9) \quad u_x^n(0,t) = \int_0^t \Phi(t,\tau)u_x^n(0,\tau)\,d\tau + \int_0^t \int_0^1 \Psi(\xi,t,\tau)F\big(\xi,\tau,u^{n-1}(\xi,\tau),u_x^{n-1}(\xi,\tau)\big)\,d\xi\,d\tau$$
$$+ \int_0^{s(t)} F\big(x,t,u^{n-1}(x,t),u_x^{n-1}(x,t)\big)\,dx - E'(t)$$

for $u_x^n(0,t)$. Since $\Phi(t,\tau)$ is a weak singular kernel, we see from the general theory of integral equations that (3.9) has a unique solution $u_x^n(0,t) \equiv \Psi^n(t)$ for given u^{n-1}. Consequently, we solve the following standard semi-linear heat equation

$$(3.10) \qquad \begin{aligned} u_t^n &= u_{xx}^n + F(x,t,u^n,u_x^n), & 0 < x < 1, \quad 0 < t \le T; \\ u^n(x,0) &= 0, & 0 < x < 1; \\ u_x^n(1,t) &= 0, & 0 < t \le T; \\ u_x^n(0,t) &= \Psi^n(t), & 0 < t \le T. \end{aligned}$$

It follows by Cannon [2] that problem (3.10) possesses a unique solution u^n which can be represented as

$$(3.11) \qquad u^n(x,t) = -2\int_0^t \theta(x,t-\tau)\Psi^n(\tau)\,d\tau$$
$$+ \int_0^t \int_0^1 N(x,\xi,t,\tau)F\big(\xi,\tau,u^n(\xi,\tau),u_x^n(\xi,\tau)\big)\,d\xi\,d\tau$$

$$(3.12) \qquad u_x^n(x,t) = -2\int_0^t \theta_x(x,t-\tau)\Psi^n(\tau)\,d\tau$$
$$+ \int_0^t \int_0^1 N_x(x,\xi,t,\tau)F\big(\xi,t,u^n(\xi,\tau),u_x^n(\xi,\tau)\big)\,d\xi\,d\tau.$$

Thus, we see that $\{u^n\}$ is well-defined for $n \ge 1$.

We shall now, first, show that there exists a function $u(x,t)$ such that

(3.13) $$u^n(x,t) \to u(x,t), \qquad u_x^n(0,t) \to u_x(0,t),$$

(3.14) $$\int_0^1 \left| u_x^n(x,t) - u_x(x,t) \right| dx \to 0$$

uniformly as $n \to \infty$, and then, employing (3.13)-(3.14) to show that

(3.15) $$u_x^n(x,t) \to u_x(x,t)$$

uniformly as $n \to \infty$. For this purpose, let $z^n = u^{n+1} - u^n$, $p^n(t) = u_x^{n+1}(0,t) - u_x^n(0,t)$. Recall that $\Psi^n(t)$ in (3.11) and (3.12) is $u_x^n(0,t)$. Then we see from (3.9)-(3.12) and standard heat kernel estimates in [2, 12] that

(3.16) $$\left| z^n(x,t) \right| \le C\sqrt{t} \|p^n\|_t + C\sqrt{t} \left\{ \|z^n(\cdot,t)\| + \sup_{0<\tau<t} \int_0^1 \left| z_x^n(x,\tau) \right| dx \right\},$$

(3.17) $$\int_0^1 \left| z_x^n(x,t) \right| dx \le C\sqrt{t} \|p^n\|_t + C\sqrt{t} \left\{ \|z^n(\cdot,t)\| + \sup_{0<\tau<t} \int_0^1 \left| z_x^n(x,\tau) \right| dx \right\},$$

and

(3.18) $$\left| p^n(t) \right| \le C\sqrt{t} \|p^n\|_t + C \left\{ \|z^{n-1}(\cdot,t)\| + \sup_{0<\tau<t} \int_0^1 \left| z_x^{n-1}(x,\tau) \right| dx \right\},$$

where the constant C in (3.16)-(3.18) depends upon T and L only. Set

(3.19) $$\||z^n\||_t = \|z^n(\cdot,t)\| + \sup_{0<\tau<t} \int_0^1 \left| z_x^n(x,\tau) \right| dx.$$

It is easy to see from (3.16)-(3.18) that

(3.20) $$\||z^n\||_t \le \frac{2C\sqrt{t}}{1-2C\sqrt{t}} \|p^n\|_t, \qquad \|p^n\|_t \le \frac{C}{1-C\sqrt{t}} \||z^{n-1}\||_t,$$

so that

(3.21) $$\||z^n\||_t \le \frac{2C^2\sqrt{t}}{(1-2C\sqrt{t})(1-C\sqrt{t})} \||z^{n-1}\||_t,$$

(3.22) $$\|p^n\|_t \le \frac{2C^2\sqrt{t}}{(1-2C\sqrt{t})(1-C\sqrt{t})} \|p^{n-1}\|_t.$$

We select a $\epsilon_1 = \epsilon_1(T,L) > 0$ such that

(3.23) $$0 \le \frac{2C^2\sqrt{t}}{(1-2C\sqrt{t})(1-C\sqrt{t})} \le \frac{1}{2}, \quad \text{for } 0 \le t \le \epsilon_1,$$

then (3.21)-(3.22) become

(3.24) $$\||z^n\||_{\epsilon_1} \le \frac{1}{2} \||z^{n-1}\||_{\epsilon_1}, \qquad \|p^n\|_{\epsilon_1} \le \frac{1}{2} \|p^{n-1}\|_{\epsilon_1}.$$

Thus, it follows from (3.24) that there exists a $u(x,t)$ such that (3.13)-(3.14) are valid. In order to show that (3.15) also holds for this $u(x,t)$, we let $z^{n,m} = u^n - u^m$, $p^{n,m} = u_x^n(0,t) - u_x^m(0,t)$, n, $m \ge 1$. Then, we see from (3.11)-(3.12) that

(3.25) $$\left| z^{n,m} \right| \le C\|p^{n,m}\|_t + C\sqrt{t} \left\{ \|z^{n,m}(\cdot,t)\| + \|z_x^{n,m}(\cdot,t)\| \right\}$$

(3.26) $$\left| z_x^{n,m} \right| \le C\|p^{n,m}\|_t + C\sqrt{t} \left\{ \|z^{n,m}(\cdot,t)\| + \|z_x^{n,m}(\cdot,t)\| \right\},$$

so that

$$(3.27) \qquad \|z^{n,m}(\cdot,t)\| + \|z_x^{n,m}(\cdot,t)\| \leq \frac{2C}{(1-2C\sqrt{t})}(\|p^{n,m}\|_t),$$

where the constant C in (3.25)-(3.27) depends only upon T and L. If we select a $\epsilon_2 = \epsilon_2(T,L) > 0$ such that $0 \leq 2C(1-2C\sqrt{t})^{-1} \leq 1/2$ for all $0 \leq t \leq \epsilon_2$ and let $\epsilon = \min\{\epsilon_1,\epsilon_2\}$, then since $\|p^{n,m}\|_\epsilon \to 0$ as n, $m \to \infty$, we see that (3.15) holds in $0 \leq t \leq \epsilon = \min\{\epsilon_1,\epsilon_2\}$. Thus, by letting $n \to \infty$ in (3.9)-(3.12), we see that u is a solution of (3.1)-(3.7).

For uniqueness, if u^1 and u^2 are two solutions, then it follows from the unique representation of u as given in (3.3) and from the above analysis that

$$(3.28) \qquad \|\!|\!|u^1 - u^2|\!|\!|_\epsilon \leq \frac{1}{2}\|\!|\!|u^1 - u^2|\!|\!|_\epsilon.$$

Q.E.D.

Hence, $u^1 \equiv u^2$.

Now we summarize above as follows:

Theorem 3.2. Under assumption H1, there exists a unique global solution u for problem (1.1)-(1.4).

Proof: See above Lemmas.

Q.E.D.

4. Continuous dependence upon the data. In this section we shall show that the solution of (1.1)-(1.4) is continuously dependent upon the data ϕ, g, E and F while $s = s(t)$ is fixed. First, let us show

Theorem 4.1. If u^k, $(k = 1,2)$ are solutions for (3.1) and (3.7) with the data E^k, $F^k(k = 1,2)$, then there exists a $C = C(T,L,s)$ such that

$$(4.1) \qquad \|u^1 - u^2\|_1 \leq C\{\|E^1 - E^2\|_1 + \|F^1 - F^2\|_\infty\} \text{ in } 0 \leq t \leq \epsilon$$

where ϵ is defined in Section 3.

Proof: Since u^k can be represented by (3.3) and (3.7) with associated data E^k and F^k, then the analysis in Section 3 implies that

$$(4.2) \qquad \|\!|\!|u^1 - u^2|\!|\!|_\epsilon \leq \frac{1}{2}\|\!|\!|u^1 - u^2|\!|\!|_\epsilon + C\{\|E^1 - E^2\|_1 + \|F^1 - F^2\|_\infty\},$$

$$(4.3) \qquad \|u^1 - u^2\|_1 \leq C\|u_x^1(0,t) - u_x^2(0,t)\|_t + C\{\|E^1 - E^2\|_1 + \|F^1 - F^2\|_\infty\}$$

$$(4.4) \qquad\qquad \leq C\|\!|\!|u^1 - u^2|\!|\!|_\epsilon + C\{\|U^1 - E^2\|_1 + \|F^1 - F^2\|_\infty\}, \quad 0 \leq t \leq \epsilon.$$

Q.E.D.

Thus, (4.1) follows from (4.2)-(4.3).

Now let us show

Theorem 4.2. If v^k, $(k = 1,2)$ is solution of (2.9) with the data ϕ^k and g^k, then there exists $C = C(T,L)$ such that

$$(4.4) \qquad \|v^1 - v^2\|_1 \leq C\{\|\phi^1 - \phi^2\|_1 + \|g^1 - g^2\|_\infty\}$$

and there exists $C = C(T,L,s)$ such that

$$(4.5) \qquad \left\|\frac{d}{dt}\int_0^{s(t)}(v^1 - v^2)(x,t)\,dx\right\|_1 \leq C\|v^1 - v^2\|_1.$$

Proof: The inequality (4.4) follows from standard estimates [2]. Since

$$(4.6) \qquad \frac{d}{dt}\int_0^{s(t)} v^k(x,t)\,dx = v^k(s(t),t)s'(t) + v_x^k(s(t),t) - v_x^k(0,t)$$

for $k = 1, 2$, then (4.5) is just a consequence of (4.4).

Finally, we have

Theorem 4.3. Under assumption H1, the solution of (1.1)-(1.4) is continuously dependent upon that the data while $s = s(t)$ is fixed.

Proof: Suppose $u^k (k = 1, 2)$ are solutions of (1.1)-(1.4) which correspond respectively to the data ϕ^k, g^k, E^k and $F^k (k = 1, 2)$ with a fixed $s = s(t)$. We rewrite u^k as $u^k = v^k + w^k$, where v^k and w^k are the corresponding solutions of (2.9) and (2.10), respectively. From Theorem 4.1 and Theorem 4.2 we see that

$$(4.7) \qquad \|w^1 - w^2\|_1 \leq C\{\|E^1 - E^2\|_1 + \|F^1 - F^2\|_\infty + \|v^1 - v^2\|\}$$

in $0 \leq t \leq \epsilon$ for some small $\epsilon = \epsilon(T, L)$. Thus, it follows that

$$(4.8) \qquad \|u^1 - u^2\|_1 \leq \|v^1 - v^2\|_1 + \|w^1 - w^2\|_1 \leq Cd, \text{ in } 0 \leq t \leq \epsilon,$$

where

$$(4.9) \qquad d = \|\phi^1 - \phi^2\|_1 + \|g^1 - g^2\|_\infty + \|E^1 - E^2\|_1 + \|F^1 - F^2\|_\infty.$$

For $\epsilon \leq t \leq 2\epsilon$. It is also easy to see above analysis and (2.11) that

$$(4.10) \qquad \|u^1 - u^2\|_1 \leq Cd + C\|v^1(\cdot, \epsilon) - u^2(\cdot, \epsilon)\|_1$$
$$\leq 2Cd, \text{ in } \epsilon \leq t \leq 2\epsilon.$$

Hence, Theorem 4.3 follows from repeating the above procedure finitely many times.

5. The first boundary value case. In this section we shall consider the problem (1.5)-(1.8) stated in Section 1. Now let v be the solution of

$$(5.1) \qquad \begin{array}{ll} v_t = v_{xx}, & 0 < x < 1, \quad 0 < t \leq T; \\ v(x, 0) = \phi(x), & 0 < x < 1; \\ v(1, t) = g(t), & 0 < t \leq T; \\ v_x(0, t) = \phi'(0), & 0 < t \leq T. \end{array}$$

We see from Cannon [2] that (5.1) has a unique solution v such that v, v_x are continuous in $0 \leq x \leq 1$, $0 \leq t \leq T$ and $\int_0^{s(t)} v(x, t) \, dx \in C^1[0, T]$ provided that $\phi \in C^1[0, 1]$, $g \in C^1[0, T]$ and $\phi(1) = g(0)$.

We write the solution u of (1.5)-(1.8) as $u = v + w$, where w is such that

$$(5.2) \qquad \begin{array}{ll} w_t = w_{xx} + F(x, t, v + w, v_x + w_x), & 0 < x < 1, \quad 0 < t \leq T; \\ w(x, 0) = 0, & 0 < x < 1; \\ w(1, t) = 0, & 0 < t \leq T; \\ \int_0^{s(t)} w(x, t) \, dx = E(t) - \int_0^{s(t)} v(x, t) \, dx, & 0 < t \leq T. \end{array}$$

Thus, we see from the analysis in Section 1 that if we can show (5.2) has a solution, then (1.5)-(1.8) will have a solution. The analysis in Section 2 implies that in order to show (1.5)-(1.8) has a solution it is sufficient to prove the following:

Theorem 5.1. Let s, $E \in C^1([0, T])$, F satisfy the assumption H1 and

$$(5.3) \qquad E'(0) = \int_0^{s(0)} F(x, 0, 0, 0) \, dx \quad E(0) = 0, \quad 0 < s(t) < 1,$$

then, there exists a $\epsilon = \epsilon(T, L)$ such that the problem

$$(5.4) \qquad \begin{array}{ll} u_t = u_{xx} + F(x, t, u, u_x), & 0 < x < 1, \quad 0 < t \leq T; \\ u(x, 0) = 0, & 0 < x < 1; \\ u(1, t) = 0, & 0 < t \leq T; \end{array}$$

and

$$(5.5) \qquad \int_0^{s(t)} u(x,t)\, dx = E(t), \quad 0 < t \leq T, \quad 0 < s(t) < 1$$

has a unique local solution u in $0 \leq t \leq \epsilon$.

Proof: It is easy to see as before that the condition (5.5) is equivalent to the following

$$(5.6) \qquad u_x(0,t) = u\big(s(t),t\big)s'(t) + u_x\big(s(t),t\big) - E'(t) + \int_0^{s(t)} F\big(x,t,u(x,t),u_x(x,t)\big)\, dx.$$

Now we shall eliminate the terms $u\big(s(t),t\big)$ and $u_x\big(s(t),t\big)$ in (5.6). Let u be a solution of (5.4)-(5.5), then by Cannon [2], u can be represented as

$$(5.7) \qquad u(x,t) = -2\int_0^t \theta(x,t-\tau)u_x(0,\tau)\, d\tau + 2\int_0^t \theta(1-x,t-\tau)u_x(1,\tau)\, d\tau$$
$$+ \int_0^t \int_0^1 N(x,\xi,t,\tau)F\big(\xi,\tau,u(\xi,\tau),u_\xi(\xi,\tau)\big)\, d\xi\, d\tau,$$

and it can also be represented as

$$(5.8) \qquad u(x,t) = -2\int_0^t \theta_x(x,t-\tau)u(0,\tau)\, d\tau + \int_0^t \int_0^1 G(x,\xi,t,\tau)F\big(\xi,\tau,u(\xi,\tau),u_\xi(\xi,\tau)\big)\, d\xi\, d\tau,$$

where G is the Green's function given by

$$G(x,\xi,t,\tau) = \theta(x-\xi,t-\tau) - \theta(x+\xi,t-\tau), \quad t > \tau.$$

Differentiating (5.7) with respect to x we obtain

$$(5.9) \qquad u_x(x,t) = -2\int_0^t \theta_x(x,t-\tau)u_x(0,\tau)\, d\tau + 2\int_0^t \theta_x(1-x,t-\tau)u_x(1,\tau)\, d\tau$$
$$\int_0^t \int_0^1 N_x(x,\xi,t,\tau)F\big(\xi,\tau,u(\xi,\tau),u_\xi(\xi,\tau)\big)\, d\xi\, d\tau.$$

Thus, we see that

$$(5.10) \qquad u\big(s(t),t\big) = -2\int_0^t \theta\big(s(t),t-\tau\big)u_x(0,\tau)\, d\tau + 2\int_0^t \theta\big(1-s(t),t-\tau\big)u_x(1,\tau)\, d\tau$$
$$+ \int_0^t \int_0^1 N\big(s(t),\xi,t,\tau\big)F\big(\xi,\tau,u(\xi,\tau),u_\xi(\xi,\tau)\big)\, d\xi\, d\tau$$

$$(5.11) \qquad u_x\big(s(t),t\big) = -2\int_0^t \theta_x\big(s(t),t-\tau\big)u_x(0,\tau)\, d\tau + 2\int_0^t \theta_x\big(1-s(t),t-\tau\big)u_x(1,\tau)\, d\tau$$
$$+ \int_0^t \int_0^1 N_x\big(s(t),\xi,t,\tau\big)F\big(\xi,\tau,u(\xi,\tau),u_\xi(\xi,\tau)\big)\, d\xi\, d\tau$$

$$(5.12) \qquad u(0,t) = -2\int_0^t \theta(0,t-\tau)u_x(0,\tau)\, d\tau + 2\int_0^t \theta(-1,t-\tau)u_x(1,\tau)\, d\tau$$
$$+ \int_0^t \int_0^1 N(0,\xi,t,\tau)F\big(\xi,\tau,u(\xi,\tau),u_\xi(\xi,\tau)\big)\, d\xi\, d\tau.$$

Differentiating (5.8) with respect to x and letting $x \to 1$ we see that

$$(5.13) \qquad u_x(1,t) = -2\int_0^t \theta_{xx}(1,t-\tau)u(0,\tau)\, d\tau + \int_0^t \int_0^1 G_x(1,\xi,t,\tau)F\big(\xi,\tau,u(\xi,\tau),u_\xi(\xi,\tau)\big)\, d\xi\, d\tau.$$

Hence, (5.6) can be rewritten as

$$(5.14) \quad u_x(0,t) = \int_0^t \Phi(t,\tau)u_x(0,\tau)\,d\tau + \int_0^t \Phi_1(t,\tau)u_x(1,\tau)\,d\tau - E'(t)$$

$$+ \int_0^t \int_0^t \Psi(\xi,t,\tau)F(\xi,\tau,u(\xi,\tau),u_\xi(\xi,\tau))\,d\xi\,d\tau + \int_0^{s(t)} F(x,t,u,u_x)\,dx,$$

where, Φ, Ψ are defined in Section 3 and

$$\Phi_1(t,\tau) = 2\theta(1-s(t),t-\tau) + 2\theta_x(1-s(t),t-\tau),$$

$$|\Phi_1| \leq \frac{C}{\sqrt{t-\tau}}, \quad t > \tau, \quad 0 < s(t) < 1.$$

If we set $p(t) = u_x(0,t)$, $q(t) = u_x(1,t)$ and $r(t) = u(0,t)$, then (5.7), (5.9) and (5.12)-(5.14) can be viewed as a semi-linear heat equation coupled with a Volterra system. We now shall solve this system by the method of iteration. Let $u^0 = 0$, $\{u^n, p^n, q^n, r^n\}$ is defined as follows. Take u^{n-1} as known, solve the volterra integral system (5.12)-(5.14) with u replaced by u^{n-1} to obtain $\{p^n, q^n, r^n\}$, and then, solve semi-linear heat equation initial-boundary value problem

$$(5.15) \quad \begin{aligned} u_t^n &= u_{xx}^n + F(x,t,u^n,u_x^n), & 0 < x < 1, & \quad 0 < t \leq T; \\ u^n(x,0) &= 0, & 0 < x < 1; \\ u_x^n(0,t) &= p^n(t), & u_x^n(1,t) = q^n(t), & \quad 0 < t \leq T. \end{aligned}$$

This yields u^n. By Cannon [2] this u^n can be represented by (5.7). Thus, we see that $\{u^n, p^n, q^n, r^n\}$ is well-defined and $u_x^n(x,t)$ is given by (5.9).

Let $Z^n = u^{n+1} - u^n$, $P^n = p^{n+1} - p^n$, $Q^n = q^{n+1} - q^n$ and $R^n = r^{n+1} - r^n$. Then, the analysis in Section 3 implies that there exists a C which depends only upon T and L such that

$$(5.16) \quad \|P^n\|_t + \|Q^n\|_t + \|R^n\|_t \leq \frac{3C}{1-3C\sqrt{t}}\||z^{n-1}|\|_t$$

and

$$(5.17) \quad \||Z^n|\|_t \leq \frac{2C\sqrt{t}}{1-2C\sqrt{t}}(\|P^n\|_t + \|Q^n\|_t + \|R^n\|_t).$$

If we select a small $\epsilon_1 > 0$ such that

$$(5.18) \quad 0 \leq \frac{6C^2\sqrt{t}}{(1-2C\sqrt{t})(1-3C\sqrt{t})} \leq \frac{1}{2}, \quad 0 \leq t \leq \epsilon_1,$$

then it follows that there exist $u(x,u)$, $p(t)$, $q(t)$ and $r(t)$ such that

$$(5.19) \quad u^n(x,t) \to u(x,t), \quad \int_0^1 |u_x^n(x,t) - u_x(x,t)|\,dx \to 0,$$

$$(5.20) \quad p^n(t) \to p(t), \quad q^n(t) \to q(t), \quad r^n(t) \to r(t)$$

uniformly as $n \to \infty$, and we also see from (5.19)-(5.20) and the analysis given in Section 3 that there exists an $\epsilon_2 = \epsilon_2(T,L)$ such that

$$(5.21) \quad u_x^n(x,t) \to u_x(x,t)$$

uniformly as $n \to \infty$ in $0 \leq t \leq \epsilon_2$. Let $\epsilon = \min\{\epsilon_1, \epsilon_2\}$ and $n \to \infty$. Thus, the system (5.7) and (5.12)-(5.14) has a solution $\{u,p,q,r\}$ and it is unique.

We shall show that u is a solution of (5.4) and (5.6). We know from (5.15) that $u_x(0,t) = p(t)$ and $u_x(1,t) = q(t)$. Equation (5.12) and Cannon [2] implies that $u(0,t) = r(t)$ and then it follows from (5.14) that (5.7) is satisfied. It remains to be shown that $u(1,t) = 0$. For this purpose, we let U be the unique solution of

(5.22)
$$\begin{array}{ll} U_t = U_{xx} + F(x,t,u,u_x), & 0 < x < 1, \quad 0 < t \leq T, \\ U(x,0) = 0, & 0 < x < 1, \\ U(0,t) = u(0,t), & 0 < t \leq T, \\ U(1,t) = 0, & 0 < t \leq T. \end{array}$$

Then, it can be represented by (5.8) and $U_x(1,t)$ is given by (5.13), so that $U_x(1,t) = u_x(1,t)$. Hence, $U(x,t) \equiv u(x,t)$ by uniqueness of parabolic equations. Consequently, u is the unique solution of (5.7).

The continuous dependence upon the data can be discussed as in Section 4, but we omit the details because of the similarity in the analysis. Now let us state our theorem for problem (1.5)-(1.8).

Theorem 5.2. If $\phi \in C^1([0,1])$, $s, E, g \in C^1([0,T])$ and F satisfies assumption H1 and, in addition, we assume the compatibility condition

$$\phi(1) = g(0), \qquad 0 < s(t) < 1,$$

$$E'(0) = \phi(s(0))s'(0) + \phi'(s(0)) - \phi'(0) + \int_0^{s(0)} F(x,0,\phi,\phi')\,dx,$$

then there exists a unique solution u for problem (1.5)-(1.8) which depends continuously upon the data with fixed $s = s(t)$.

References

1. J. R. CANNON, *The solution of the heat equation subject to the specification of energy*, Quart. Appl. Math., Vol. 21, No. 2 (1963), pp. 155-160.

2. J. R. CANNON, *The one dimensional heat equation*, in the Encyclopedia of Mathematics and its Applications, Vol. 23, Addison- Wesley, Menlo Park, Calif., 1984.

3. J. R. CANNON and J. VAN DER HOEK, *The existence of and a continuous dependence result for the solution of the heat equation subject to the specification of energy*, Boll. Uni. Math. Ital. Suppl. 1 (1981), pp. 253-282.

4. J. R. CANNON and J. VAN DER HOEK, *An implicit finite difference scheme for the diffusion of mass in a portion of the domain*, Numerical Solutions of Partial Differential Equations (J. Noye, ed.), North-Holland, Amsterdam, 1982, pp. 527-539.

5. J. R. CANNON and J. VAN DER HOEK, *The one-phase Stefan problem subject to the specification of energy*, J. Math. Anal. Appl., Vol. 86 (1982), pp. 281-291.

6. J. R. CANNON and J. VAN DER HOEK, *The one-dimensional two- phase Stefan problem subject to the specification of energy*, Annali di Mat. Pura ed Applicata (IV), Vol. CXXX (1982), pp. 385-398.

7. J. R. CANNON and J. VAN DER HOEK, *Diffusion subject to the specification of mass*, J. Math. Anal. Appl., Vol. 115, No. 2 (1986), pp. 517-529.

8. J. R. CANNON, SALVADOR PEREZ ESTEVA and J. VAN DER HOEK, *A Galerkin procedure for the diffusion equation subject to the specification of mass*, SIAM J. Numer. Anal., 24 (1987), pp. 499- 515.

9. W. A. DAY, *Existence of a property of solutions of the heat equation to linear thermoelasticity and other theories*, Quart. Appl. Math., 40 (1982), pp. 319-330.

10. W. A. DAY, *A decreasing property of solutions of a parabolic equation with applications to thermoelasticity*, Quart. Appl. Math., 41 (1983), pp. 468-475.

11. A. FRIEDMAN, *Monotonic decay of solutions of parabolic equations with nonlocal boundary conditions*, Quart. Appl. Math., Vol. XLIV, No. 3 (1986), pp. 401-407.

12. A. FRIEDMAN, *Partial Differential Equations of Parabolic Type*, Prentice-Hall, Inc., 1964.

13. N. I. IONKIN, *Solution of a boundary-value problem in heat conduction with a nonclassical boundary condition*, Diff. Eqs., Vol. 13, No. 2 (1977), pp. 294-304.

14. N. I. IONKIN and E. I. MOISEEV, *A problem for the heat transfer equation with two-point boundary condition*, Diff. Eqs., Vol. 15, No. 7 (1979), pp. 1284-1295.

15. L. I. KAMYIN, *A boundary value problem in the theory of heat conduction with a nonclassical boundary condition*, U.S.S.R. Comput. Maths. Math. Phys., Vol. 4, No. 6 (1964), pp. 33-59.

16. BERNHARD KAWOHL, Remark on a paper by W. A. DAY, *A maximum principle under nonlocal boundary conditions*, Quart. Appl. Math., Vol. XLIV, No. 4 (1987), pp. 751-752.

17. O. A. LADYZENSKAJA, V. A. SOLONNIKOV and N. N. URAL'CEVA, *Linear and quasilinear equations of parabolic type*, Amer. Math. Soc. Trans., Vol. 23 (1968), A.M.S., Providence, R.I.

18. V. L. MAKAROV and D. T. KULYEV, *Solution of a boundary- value problem for a quasilinear parabolic equation with a nonclassical boundary condition*, Diff. Eqs., Vol. 21, No. 2 (1985), pp. 296-305.

19. N. I. YURCHUK, *Mixed problem with an integral condition for certain parabolic equations*, Diff. Eqs., Vol. 22, No. 12 (1986), pp. 2117-2126.

CHAPTER 4

Development of Complex Reaction Front Morphologies
Through Nonlinear Fluid-Rock Interaction: Modeling,
Asymptotic and Numerical Studies**

W. Chen*
P. Ortoleva*

Abstract. The morphology of reaction fronts in porous media is analyzed numerically in the context of a specific example motivated by geochemical applications. The numerical simulations are greatly aided by the reduction of the reaction-transport equations via a two fold multiple time scale analysis that takes advantage of extreme values of the parameters appropriate to most natural geochemical systems.

The main result is the illustration of a hierarchy of reaction front morphologies that unfold when the rate of fluid injection into the system increases. We demonstrate c-shaped, scalloped, branching-tree and other forms. Finger tip splitting instabilities and temporally oscillatory states involving finger meandering and budding are also observed.

INTRODUCTION. Moving fronts develop in porous media when fluids injected into them react with components of the matrix. The coupling of flow with reactions affecting porosity and permeability can lead to the morphological instability of high symmetry (planar or spherical) fronts to the formation of scalloping or fingering[1-10] as suggested in Fig. 1. Here we investigate the complex morphologies of these fronts that may develop when these systems are driven very far from the primary bifurcation point wherein the planar front just loses stability.

---

*Departments of Chemistry and Geological Sciences, Indiana University, Bloomington, IN 47405.
**Research supported in part by a contract with the Gas Research Institute, and the U.S. Department of Energy, Basic Energy Sciences Division.

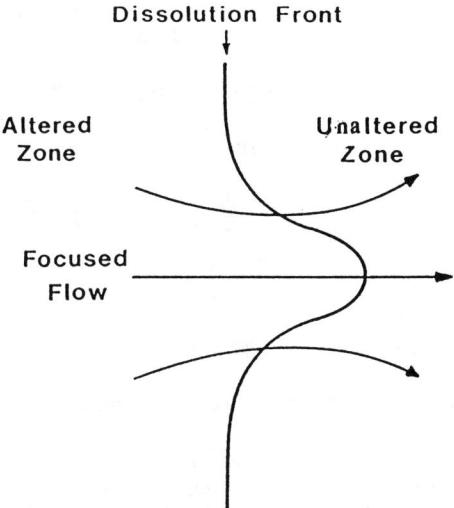

Figure 1. Schematic view of nonplanar reaction front illustrating flow self-focusing.

Reaction fronts play key roles in natural and engineered geological systems (see Refs. 1-10 and citations therein). Fronts are an important aspect of the development of ore bodies (e.g. for uranium and copper), diagenetic alteration leading to the production or destruction of good petroleum reservoir rock, and possibly to the development of diagenetic petroleum traps. Enhanced oil recovery and well stimulation or completion also can yield reaction fronts.

The basic morphological instability is illustrated in Fig. 1. Consider a fluid injected into a medium consisting of an insoluble matrix and a soluble component. In time a dissolution front develops wherein the soluble component is eliminated upstream of the front and the fluid is saturated with respect to the soluble mineral downstream from it. Within the front the soluble mineral is dissolving into molecules in the pore fluid and the concentrations of these molecules attains saturation as the fluid passes through the reaction zone. But as the mineral dissolves, the permeability and porosity increase. Thus within a bump in the reaction front (as in Fig. 1) the permeability and porosity is larger than in regions adjacent to the bump in the unaltered zone. This tends to focus the flow to the bump. This augmented flow-through increases the rate of advancement of the tip of the bump, completing a "flow self-focusing instability".[1-3] Note that this phenomenon is distinct from the Saffman-Taylor instability[11,12] arising from the viscosity difference between two fluids when one fluid (the less viscous) pushes the other out of the porous medium. In that case the front moves at the speed of the fluid. In contrast, the reaction fronts of interest here move orders of magnitude slower than the fluid speed due to the very low solubility of most minerals.

In earlier work[1-3] a simple dissolution front problem describing the model

$$\text{solid} \rightleftarrows \text{pore fluid solute} \tag{1}$$

was set forth. The planar front was shown to be linearly unstable in a finite width channel beyond a critical imposed fluid flow rate. Numerical simulations were used to illustrate the formation of a dissolution finger that developed beyond this primary bifurcation point.[1] A weakly nonlinear analysis was also carried out to determine the amplitude (length) of the finger for near critical flow speeds.[2]

Our objectives here are twofold:

* extend the analysis to more complex (and thereby realistic) systems; and
* explore reaction front morphologies far above the primary bifurcation point.

The first objective is met by resolving a number of asymptotic and computational points while the latter results in the demonstration of the richness of the system to display a range of spatially and temporally complex behaviors.

REACTION-TRANSPORT MODEL. The mathematical model to be used is based on two types of equations –

* conservation of solute and solvent mass
* grain growth/dissolution.

Let us review these equations in some generality so as to comment on their structure and potential for nonlinear dynamical behavior. In the next section we examine their asymptotic behavior in some geologically relevant limits.

Description. The medium is specified via a macroscopic description in terms of a set of variables specifying the local state of the medium:

N = number of pore fluid species

M = number of minerals

α = pore fluid solute species label ($=1,2,...N$)

i = mineral label ($=1,2,...M$)

c_α= concentration of solute α in pore fluid (assumed constant within a pore)

ξ_i= set of variables describing the disposition of mineral i grains.

The ξ_i specify the size, shape, orientation and other variables needed to characterize the "typical" mineral i grain and its relation to its neighbors (i.e. contact areas and shape). Other variables needed to completely specify the local state of the rock are the temperature, fluid pressure, and stress (applied to each macro-volume element of the medium).

Less fundamental quantities are expressible in terms of the above variables. These include the porosity ϕ, the number of grains n_i of i mineral grains per system volume and the velocity(ies) of the various fluid phases (water, oil, natural gas). Two such relations are quite important. For many applications we can neglect the dependence of grain volume on stress and temperature. Under these conditions we take $V_i(\xi_i)$ to be the volume of the typical mineral i grain. Then we have

$$\phi + \sum_{i=1}^{M} n_i V_i = 1 , \tag{2}$$

insuring that all space is filled with either pores or mineral grains. If no grains are nucleated or destroyed, then kinematics implies

$$\frac{\partial n_i}{\partial t} = -\vec{\nabla} \cdot (n_i \vec{u}) \tag{3}$$

where $\vec{u}$ is the velocity of deformation of the porous matrix.

Texture Dynamics. The evolution of the texture $\underline{\Theta}$,

$$\underline{\Theta} = \{\xi_1, \xi_2, ... \xi_M\} \tag{4}$$

is dictated by the kinetics of grain growth, dissolution and (possible) mechanical deformation. Let ξ_{im} be the m-th variable used to characterize a mineral i grain and let G_{im} be its rate of change due to mineral grain growth, dissolution or mechanical deformation. For a small time increment δt we have

$$\xi_{im}(\vec{r}, t+\delta t) = \xi_{im}(\vec{r}-\vec{u}\delta t, t) + \delta t G_{im} \tag{5}$$

for spatial position $\vec{r}$ and time t. Because δt is arbitrary (but small) this implies

$$\frac{\partial \xi_{im}}{\partial t} = -\vec{u} \cdot \vec{\nabla} \xi_{im} + \sum_k \psi_{ik} G_{imk} \tag{6}$$

where ψ_{ik} is a stoichiometric coefficient for the growth of an i formula unit from process k. Combining these equations with (3) for n_i yields a constraint on $\vec{u}$:

$$\frac{\partial}{\partial t}(1-\phi) + \vec{\nabla} \cdot [(1-\phi)\vec{u}] = \sum_{imk} \psi_{ik} n_i V_{im} G_{imk}$$

$$V_{im} \equiv \partial V_i / \partial \xi_{im} \; . \tag{7}$$

This provides a constraint on the deformation velocity. For example, in a low porosity rock ($\phi \ll 1$) we obtain

$$\vec{\nabla} \cdot \vec{u} = \sum_{imk} \psi_{ik} n_i V_{im} G_{imk} \; (\phi=0) \; . \tag{8}$$

This is an important result for deep rocks (such as under metamorphic conditions).[9-13] In the general case ($\phi \neq 0$) (7) still provides a constraint on $\vec{u}$ once the dependence of ϕ and the n_i, V_i and G_{imk} on the fundamental descriptive variables is specified. The above constraints replace the simpler divergence free condition $\vec{\nabla} \cdot \vec{u} = 0$ imposed on nonreacting deforming incompressible polycrystalline media.

Conservation of Solute and Solvent Mass. Conservation of mass for the pore fluid species gives a set of reaction-transport equations. For a single phase pore fluid we have

$$\frac{\partial \phi c_\alpha}{\partial t} = -\vec{\nabla} \cdot \vec{J}_\alpha + \phi \sum_{k=1}^{N_f} \nu_{\alpha k} \frac{W_k}{\varepsilon} + \sum_{imnk} \mu_{\alpha ik} \rho_i n_i V_{im} G_{imk} \tag{9}$$

where $\vec{J}_\alpha$ is the flux of α, $\nu_{\alpha k}$ and $\mu_{\alpha ik}$ are stoichiometric coefficients, W_k is the rate of the k-th of N_f reactions taking place in the pore fluid and ρ_i is the molar density of mineral i, $V_{im} = \partial V_i / \partial \xi_{im}$. In the above we have assumed that the rate of change of grain volume due to thermal, stress, defect production or any other nonreactive process is negligible. The factor ε in (9) is included to emphasize the fact that in many geological contexts the rate coefficients associated with fluid phase reactions are large and hence these processes are maintained near equilibrium.

Phenomenological Laws. The above equations must be supplemented by phenomenological laws relating the fluxes and reaction rates to the descriptive variables. For example, the flux $\vec{J}_\alpha$ is typically taken to be of the form

$$\vec{J}_\alpha = -\phi D \vec{\nabla} c_\alpha + (\vec{u}+\vec{v}) \phi c_\alpha \tag{10}$$

where $\vec{v}$ is the fluid velocity (relative to the solid matrix). Darcy's law yields

$$\vec{v} = -\kappa [\vec{\nabla} p + \rho^m g \vec{z}] \; , \tag{11}$$

where ρ^m is the mass density of the fluid, g is the acceleration of gravity, $\vec{z}$ is a unit upward-pointing vector and κ is the (texture dependent) permeability divided by fluid viscosity.

Further relations are needed to obtain the dependence of D and κ on texture (and possibly on $\underline{c}$,p,T), as well as for ρ^m on $\underline{c}$,p,T. A further assumption is needed to obtain the fluid pressure p. In the case of a rock with pores filled with a single fluid, this is taken to be the equation of state; $p = p(\underline{c},T)$. Most commonly for relatively incompressible fluids the equation of conservation of mass for the solvent (labeled say species $\alpha=1$) is written

$$c_1 = c_1(c_2,c_3,\ldots c_N,p,T).$$ (12)

This is placed in the conservation of mass relation (9) to obtain an equation for the pressure. Finally the rates W_k and G_{imk} are dependent on $\underline{c}$ and in many cases the p,T and (for the G's) Θ and rock stress. With these many cross dependencies the fluid-rock system evolves via a complex, nonlinear, coupled dynamics.

A Highly Coupled Nonlinear Dynamical System. These coupled processes allow fluid-rock systems to display a richness of nonlinear dynamical responses when they are maintained out of equilibrium. In the present work we explore the consequences of the coupling of flow with mineral reactions via the Θ-dependence of the permeability. In a companion paper by Dewers and Ortoleva the effect of stress on reaction and, in turn, of rock mechanical properties on stress allows for symmetry breaking instabilities and temporal oscillatory dynamics as sedimentary rocks become buried kilometers beneath the surface over geologic time. These two examples serve to demonstrate that the many couplings through the transport and reaction rate laws allows for a number of feedback loops that make fluid-rock interaction systems a very promising area for future work in modeling, analysis and numerical simulation.

ASYMPTOTIC REDUCTION AND SIMPLIFICATION. A number of features of natural geochemical systems imply that the above reaction-transport model has limiting or asymptotic behavior with respect to certain system parameters. As they pertain directly to the phenomena of reaction front morphology to be considered below, let us review some of them briefly here.

Fast Fluid Phase Reactions. The fast fluid phase reaction limit is conveniently carried out by examining the asymptotic behavior ($\varepsilon \to 0$) of (9). The result is a set of reduced reaction-transport equations with subsidiary equilibrium conditions

$$W_k(\underline{c},p,T) = 0, \quad k=1,2,\ldots N_f.$$ (13)

While there are N_f fluid phase reactions, not all may be thermodynamically distinct. Thus if N_r of the N_f reactions are redundant, then only $N_f^* = N_f - N_r$ of the conditions (13) are actually independent.

To obtain the reduced set of reaction-transport equations, we introduce a set of row vectors $\underline{a}^{(\lambda)}\{a_1^{(\lambda)},a_2^{(\lambda)},\ldots a_N^{(\lambda)}\}$ for $\lambda=1,2,\ldots N-N_f^*$ such that the $\underline{a}^{(\lambda)}$ are orthogonal to the column-vectors corresponding to the columns of the stoichiometric matrix $\nu_{\alpha k}$. There are in fact N_f^* linearly independent such ν-column vectors. Thus

$$\sum_{\alpha=1}^{N} a_\alpha^{(\lambda)} \nu_{\alpha k} = 0 \text{ for all } \lambda,k .$$ (14)

Multiplying both sides of (9) by $a_\alpha^{(\lambda)}$, summing over all α and using the property (14) we obtain

$$\frac{\partial \phi \chi^{(\lambda)}}{\partial t} = -\vec{\nabla} \cdot \vec{j}^{(\lambda)} + \sum_{imkn} \omega_{imk}^{(\lambda)} n_i \rho_i V_{im} G_{imk} \tag{15}$$

$$\begin{Bmatrix} \chi^{(\lambda)} \\[2mm] \vec{j}^{(\lambda)} \\[2mm] \omega_{ik}^{(\lambda)} \end{Bmatrix} = \sum_{\alpha=1}^{N} a_\alpha^{(\lambda)} \begin{Bmatrix} c_\alpha \\[2mm] \vec{J}_\alpha \\[2mm] \mu_{\alpha ik} \end{Bmatrix} . \tag{16, 17, 18}$$

Note that as $\varepsilon \to 0$, (15) is well behaved and that in this limit (9) implies (13). The above conclusions are predicated upon the assumptions that

* the c_α are well behaved in space and time as $\varepsilon \to 0$ and
* similarly for $\partial c_\alpha / \partial t$ and the first and second spatial derivatives of the c_α.

That solutions of this type exist is only true if the equilibrium problem (13) yields a unique solution (once supplemented by $N-N_f^*$ auxiliary conditions). This is generally believed to be the case (see however Ref. 16 in the context of plasmas). If multiple solutions occur then compositional shocks may develop and propogate (see Refs. 17-20 and citations therein).

Solid Density Asymptotics. In many problems a typical solute concentration, $\bar{c}$, is much less than a typical solid molar density $\bar{\rho}$:

$$\delta \equiv \bar{c}/\bar{\rho} \ll 1 . \tag{19}$$

Furthermore, the mineral process rates G_{imk} are typically slow. For example, they typically are inversely proportional to the solid molar density ρ_i and are proportional to at least one factor of a fluid phase solute concentration. Thus we expect G_{imk} is of order δ and write

$$G_{imk} = \delta G'_{imk} . \tag{20}$$

Finally, the rock deformation rate is often due to mineral reactions leading to compaction and hence $\vec{u}$ is expected to be linearly related to some of the G_{imk}. Hence $\vec{u}$ also should scale as δ. With this we write

$$\vec{u} = \delta \vec{u}' . \tag{21}$$

If we drop the primes, (8) then attains its original form. However, the solute dynamic (15) is modified. Putting

$$c_\alpha = \bar{c} c'_\alpha, \quad \vec{J}_\alpha = \bar{c} \vec{J}'_\alpha \tag{22}$$

and (21) into (15) and taking $\delta \to 0$ yields, upon dropping the primes,

$$\vec{\nabla} \cdot \vec{j}^{(\lambda)} = \sum_{imk} \omega_{imk}^{(\lambda)} n_i \rho_i G_{imk} . \tag{23}$$

This is a steady state equation of (15). With this all system dynamics is carried by the texture evolution equations.

A HIERARCHY OF REACTION FRONT MORPHOLOGIES IN CALCITE CEMENTED SANDSTONES.

Calcite Dissolution Front Model. As our objective here is to study the morphologies of reaction fronts we adopt the simplest realistic geochemical model. The chemistry chosen is that for calcite while the textural model is a mean spherical picture as defined below. The model is then simulated numerically to explore the hierarchy of morphological transitions of a calcite dissolution front in a calcite cemented sandstone.

For a textural model we choose the mean spherical description whereby a grain of mineral i is represented by an effective sphere of radius R_i such that $V_i = 4\pi R_i^3/3$ (essentially by definition of R_i). The matrix is assumed not to deform so that $\vec{u} = \vec{0}$. From (3) this implies that n_i (the number of i grains per rock volume) is a constant in time. In the mean spherical model n_i must be specified. However ϕ is given by (2). The area factor V_{im} in (9) is simply $4\pi_i^2$, valid only for porous rocks where the grain-grain contact area is small relative to the area of a grain in contact with the pore fluid. Note that the m-sum in (9) collapses to a single term. Thus the texture dynamics (6) reduces to $\partial R_i/\partial t = \sum_k \psi_{ik}G_{ik}$ and 'if there is only one reaction per mineral, the k sum collapses to a single term.

As a working example, we consider the alteration of a calcite cemented sandstone subject to an infiltration of fluids undersaturated with respect to the cements. The chemical model adopted is[23]

$$\text{calcite} \rightleftarrows Ca^{2+} + CO_3^{2-} \tag{24}$$

$$HCO_3^- \rightleftarrows H^+ + CO_3^{2-} \tag{25}$$

$$H_2CO_3 \rightleftarrows H^+ + HCO_3^- \tag{26}$$

$$CaCO_3^0 \rightleftarrows Ca^{2+} + CO_3^{2-} \tag{27}$$

$$H_2O \rightleftarrows H^+ + OH^- . \tag{28}$$

This model appears to capture the majority speciation in nonsaline natural near-surface systems.

The central feedback that allows for the destabilization of the planar calcite dissolution front is through the dependence of the Darcy permeability on the amount of soluble component (calcite cement) within the medium. In the present study we adapt the phenomenology described by the Fair-Hatch equation:[24]

$$\kappa = \kappa_0 \phi^3/F \tag{29}$$

$\kappa_0 = (J\mu_w\Theta^2)^{-1}$ is a multiple factor, J (≈ 5) a packing factor, Θ (≈ 6 for spherical grains) a geometric factor, μ_w ($\sim .001 - .01$ poise) the water viscosity. The factor F is defined by

$$F = \sum_{i=1}^{M} \phi_i/L_i , \tag{30}$$

for the two mineral (M=2, quartz and calcite) case of interest. Here L_i is the cube root of the volume of a grain of mineral i, and ϕ_i is the volume fraction of mineral i. The diffusion coefficient D is assumed to be given by

$$D = D_0 \phi^2 \tag{31}$$

where D_0 is a multiple factor that is approximately equal to a diffusion coefficient in water.

Other reaction and transport laws adopted were

$$G_{cal} = k \left[C_{Ca^{2+}} C_{CO_3^{2-}} - K \right] \qquad (32)$$

for constants k and K. Values used are summarized in the Appendix.

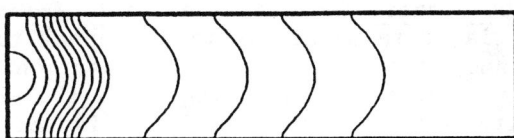

Figure 2. Temporal development of a steady finger-shaped dissolution front for a system of 7.5×2 cm; obtained by using a grid of 76×21. The time interval between adjacent contours is 10^8 sec for the first 9 contours and 8×10^8 sec for the last 5 contours. The imposed flow along the horizontal direction (specified at the outlet) is 1.4×10^{-3} cm/sec.

Genesis of A Simple Reaction Front Finger. The temporal development of the flow self-focusing induced instability of a planar dissolution front is illustrated in Fig. 2. There, a steady finger shaped dissolution front is observed to develop due to the influx of calcite undersaturated fluid into a calcite cemented sandstone. Fig. 2 (as well as all the other figures to be shown) gives the location of a reaction front at a sequence of times. Each curve represents the locus of points at one instant of time wherein the porosity value is half-way between its original value (chosen to be 10%) and that after all calcite has disappeared (chosen to be 20%). The inlet fluid, under a constant pressure, flows into the domain from the left-hand side wall. It flows out of the right-hand side wall with a constant imposed velocity along the horizontal direction. Therefore, the dissolution front moves to the right with time. The left-most curve in the figure gives the initial location of the reaction front; the first contour to its right represents the dissolution front at the next time displayed, etc. The time increment between successive reaction front locations is held to be a constant for each figure unless explicitly noted in the figure captions. Reaction rate coefficients, diffusion and permeability data used are summarized in the Appendix.

At time zero, the simulation domain in Fig. 2 is uniformly cemented with calcite except for the small high porosity "bump" located just to the right of the inlet wall. There the permeability and porosity were somewhat higher than in the unaltered rock. This small nonuniformity is amplified into a C-shaped reaction front during the dissolution process. The increase in permeability induced by calcite dissolution allows for the focusing of flow to the tip of the "dissolution finger" that sustains the nonplanarity. Diffusion of Ca^{2+} and CO_3^{2-} from the sides of the finger prevents the elongation from continuing indefinitely. These two effects tend to balance each other and the fingered reaction front eventually reaches a "steady state" whereby the front advances down stream with constant shape and velocity.

Flow self-focusing can only sustain a dissolution finger when the inlet flow speed exceeds a critical value. This value depends on inlet composition, initial calcite and inert quartz matrix volume fraction, and the channel width. The purpose of this communication is to show that this fingering of the calcite dissolution front is just one of a rich class of reaction front morphologies that may develop when the inlet flow speed or the width of the flow channel are increased.

Stability of the Finger. The reaction front of Fig. 2 developed for a channel with initial data symmetric with respect to the central line dividing its top and bottom halves. The question arises as to the stability of such a symmetric steady finger - i.e., what will happen if the initial porosity perturbation "bump" is displaced from the center of the inlet wall?

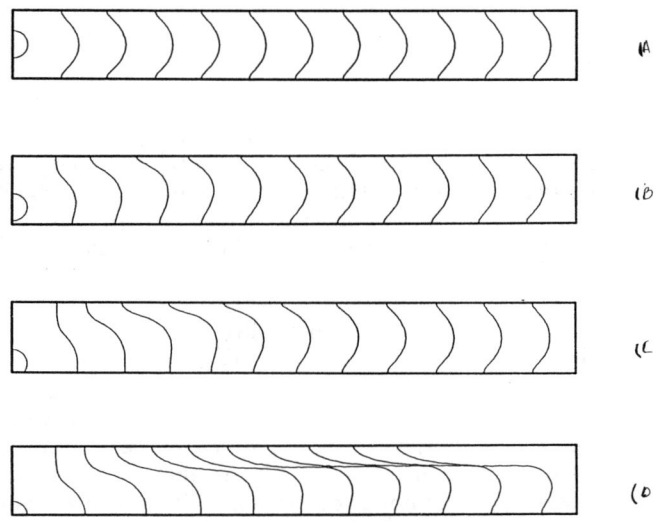

Figure 3. Stability of the finger due to symmetry breaking perturbation. The center of the initial porosity bump is placed on the inlet wall at (a) $y_c = (1/2)y_L$, (b) $y_c = (1/4)y_L$, (c) $y_c = (1/8)y_L$, (d) $y_c = 0$; $y_L = 2$ cm is the height of the domain. Note the ultimate reaction fronts of (a) through (c) are identical although the initial perturbation are different, showing that the symmetric reaction front (a) is stable with respect to symmetry breaking. Frame (d) illustrates that the existence of the asymmetric state when the initial perturbation is sufficiently away from the symmetry points. The size of the domain is 15 × 2 cm. The time interval between the adjacent contours is 10^9 sec. The imposed flow along the horizontal direction (specified at the outlet) is the same as that of Figure 2.

A series of simulations was carried out to answer this question, and the results are shown in Fig. 3. The simulation domain for Fig. 3(a) is exactly the same as that of Fig. 2 except the domain is twice as long for 3(a). Simulations shown in Figs. 3(a)-(d) only differ in the location of the initial perturbation at the left. We see that while the fingered reaction front develops and moves down stream, the finger tip gradually shifts toward the central line if the initial bump is off that line. The finger tips of 3(b) and 3(c) reach the central line after long enough time advancement and the shape of the reaction front becomes identical to that of 3(a). It is not clear from 3(d) that the finger tip would reach the central line. However another simulation for a domain much longer than 3(d) shows no further shift of the finger tip toward the central line before a steady state is reached. Thus it appears that the asymmetric state of frame (d) is a distinct steady front solution that can exist in a range of conditions that overlaps with that for the symmetric finger of frame (a). There can thus be multiple stable steady reaction front morphologies each with a finite basin of attraction.

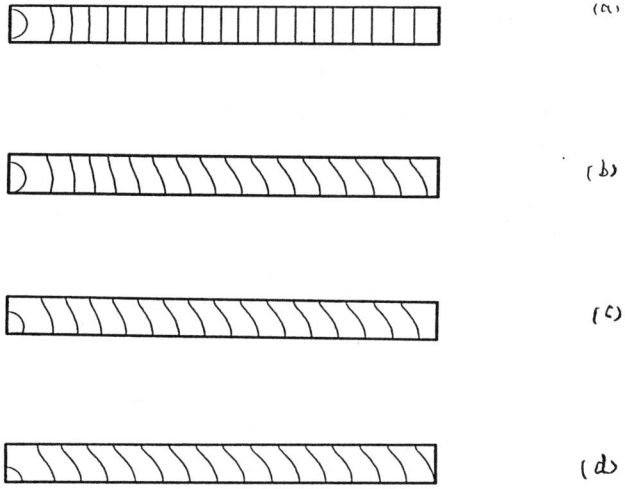

Figure 4. Evolution of reaction front in narrow domain demonstrates that the asymmetry in the placement of the initial perturbation causes the front to evolve to a stable half-finger state, as in (d). The domain used are half as wide as those in Figure 3. The center of the initial porosity bump is placed on the inlet wall at: (a)$y_c = (1/2)y_L$, (b) $y_c = (1/4)y_L$, (c) $y_c = (1/8)y_L$, (d) $y_c = 0$; $y_L = 1$ cm is the height of the domain. The size of the domain is 10 X 1 cm. The time interval between adjacent contours is .5 X 10^9 sec. The imposed flow along the horizontal direction (specified at the outlet) is the same as that of Figures 2 and 3.

In the earlier studies of Refs. 1-3 it was shown analytically in the limit of fast mineral reactions (K large in (32)) that the first nonplanar front to bifurcate from the planar solution is the "half-finger" as for example seen in frame (d) of Fig. 4. The domains for Fig. 4 are only half as wide as those for Fig. 3, while the other parameters used are exactly the same. However, unlike 3(b) and 3(c), 4(b) and 4(c) show the shift of finger tips toward the bottom boundary of the domain. After long times, the shape of the reaction fronts become identical to that of Fig. 4(d) in which the initial bump is put at the lower-left corner. The half-finger located at the bottom boundary is a stable solution with a finite basin of attraction. Note the reaction front for 4(a) is planar because the inlet fluid speed is lower than the bifurcation point for the full finger and the initial data was symmetric about the horizontal line dividing the system in half.

Figs. 3 and 4 are good examples of the competition between tendencies to evolve distinct reaction front morphologies. As a result of this competition, the finger may shift along the direction perpendicular to the overall fluid flow while the fingers develop and move down stream. For given initial data, the ultimate position of the finger depends on the basin of attraction in which the initial data resides.

Finger Tip Splitting. While flow self-focusing leads to the destabilization of a planar calcite dissolution front and the formation of a fingered front, it can destabilize a finger as well to form a pair or even a triplet of fingers. This phenomenon is termed "tip splitting".

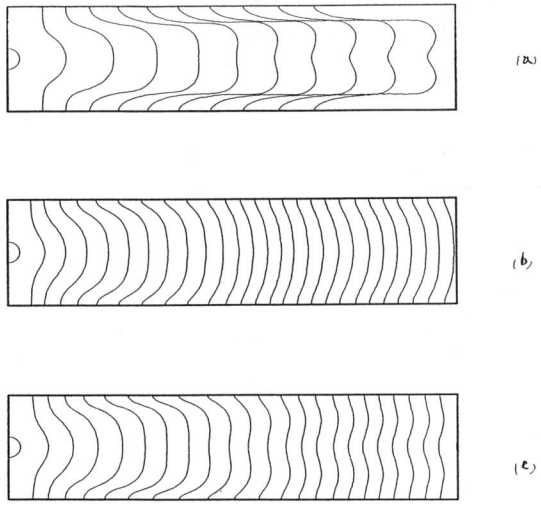

Figure 5. Finger tip splitting phenomenon whereby the reaction front divides into two branches. The imposed flow along the horizontal direction (specified at the outlet) has to be higher then the "tip splitting point" before the phenomenon is observed. The imposed flow speeds are: (a) $V_0 = 1.4 \times 10^{-3}$ cm/sec, (b) $V_0 = 8.9 \times 10^{-4}$ cm/sec, lower than the tip splitting point, (c) $V_0 = 1.0 \times 10^{-3}$ cm/sec, slightly higher than the tip splitting point. The domain is twice as wide as that of Figure 3. Its size is 15 X 4 cm. The time interval between adjacent contours is 10^9 sec.

Fig. 3(d) illustrates tip splitting. If we combine Fig. 3(d) with its mirror image obtained through a reflection across its bottom boundary, we obtain Fig. 5(a). In other words, 5(a) is a symmetric state for a system twice as wide as that of 3 with all the other system parameters the same. A comparison of the simulations 3(a), 4(a) and 5(a), different only by their domain width, demonstrating that a planar reaction front is more unstable in a wider domain under the same inlet fluid speed and other parameters.

The channel has to be wide enough in order to support a fingered reaction front or else diffusion will drive the front to planarity. With increase of channel width, it will first become possible for the system to support a half-finger state. Then it will support a full-finger state as well as the half-finger state.

The speed of the inlet flow is another factor that affects the stability of nonplanar front morphology. Fig. 5(b) shows a system exactly the same as that of 5(a) except the inlet flow speed is reduced to half of that for 5(a). We observe that only a small amplitude finger develops whose tip does not split. Obviously, there exists another critical inlet fluid speed, denoted the "tip splitting point", below which tip splitting will not occur. It is rather interesting to observe that when the inlet flow speed exceeds the tip splitting point by just a small amount, the finger will first elongate, then it will start to split as in Fig. 5. The amplitude of the deviation from planarity ultimately becomes smaller than that for the same system under a slower inlet flow that supports a single elongate finger. This is shown in Fig. 5(c) for which the inlet flow is faster than that for 5(b). and just above the tip splitting point. The stable steady state observed here is termed "double finger" while the state observed in 5(b) is "single finger".

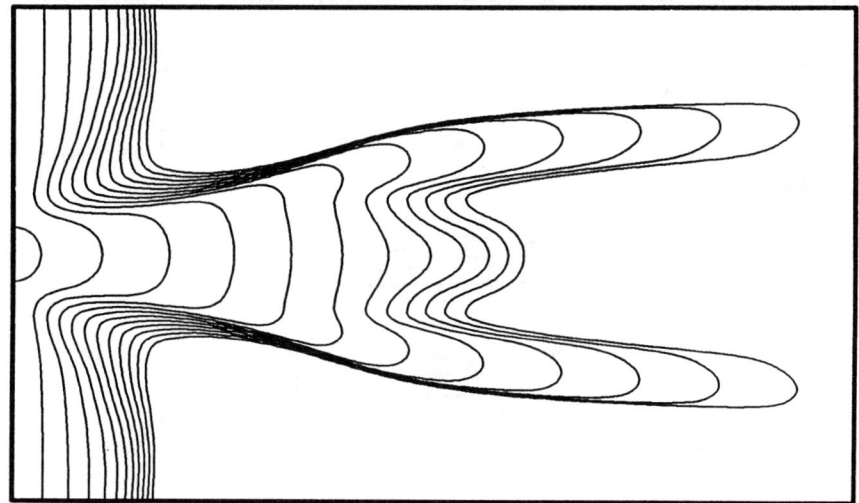

Figure 6. Triplet finger reaction front. The domain is 2 times wider than that of Figure 3 (12 X 120 cm). The imposed flow along the horizontal direction (specified at the outlet) is 5.3 X 10^{-3} cm/sec which is much faster than that used for Figure 5. The time interval between adjacent contours is 5 X 10^7 sec.

Based on the above discussions, we expect that for a wider domain and/or faster inlet flow, the system will become even more unstable in that the finger tip may split again and result in increasing complexity of the reaction front morphology. This is demonstrated in Figs. 6 and 7.

In Fig. 6 the temporal development of the half-height contour line shows the evolution of a reaction front starting from a small centered perturbation at the inlet. The initial disturbance elongates into a finger. The finger broadens and flattens. The corners attract more fluid through them and develop into a pair of fingers. Note that a central finger also develops although it is much weaker than the other two that originated from the corners.

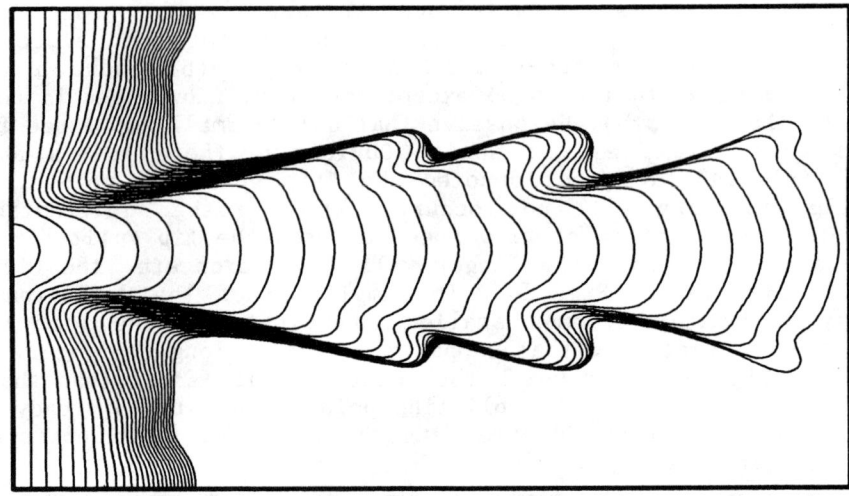

Figure 7. Small side fingers or "buds" emerge periodically in time in a frame moving with the advancing main shoot. The imposed flow is $V_0 = 7.2 \times 10^{-3}$ cm/sec, faster than that of Figure 6. The time interval between adjacent contours is 2×10^7 sec. All the other parameters used are the same as that of Figure 6.

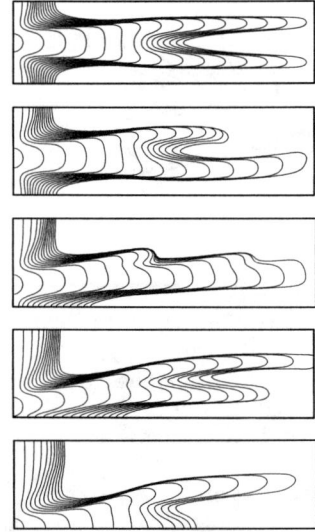

Figure 8. A host of multiple fingered front may be initiated with different initial data (here the position of the initial perturbation at the left) in a fast flow system with enough width. The center of the initial porosity bump is placed on the inlet wall at: (a) $y_c = (1/2)y_L$, (b) $y_c = (5/12)y_L$, (c) $y_c = (1/4)y_L$, (d) $y_c = (1/12)y_L$, (e) $y_c = 0$; $y_L = 6$ cm. The profile of the five dissolution fronts are substantially different from one another, showing the double fingers observed here are unstable with respect to the symmetry breaking of the system. The size of the domain is 20 X 6 cm. The time interval between adjacent contours is 5×10^7 sec. The imposed flow is $V_0 = 5.3 \times 10^{-3}$ cm/sec.

In Fig. 7 the inlet fluid flow is faster than that of Fig. 6, while all the other parameters are held the same. In this case the central finger is much stronger than putative side fingers. We term the development of small side fingers "budding". We see the buds arise periodically in time in a frame moving with the advancing central shoot.   This suggests that there may be a critical value of inlet velocity at which a steady nonplanar state undergoes a "Hopf bifurcation"[23] to a temporally oscillatory budding state. While our numerical results did not rigorously confirm this, it appears that the dynamics illustrated in Fig. 7 is time-periodic in a reference frame moving with the average motion of the front.

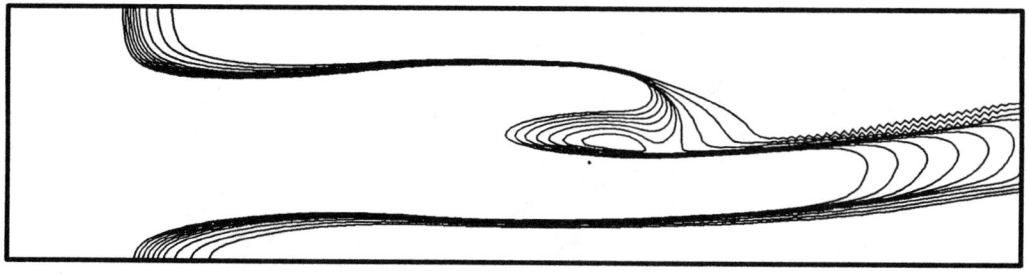

Figure 9. Reinjection of a small branch into the main shoot of the dissolution front. The system is exactly the same as in 6(b) except that it is longer by 15%; the time interval between adjacent contours is 2 X 10^7 sec and the first left-most contour line is at time = 6.5 X 10^8 sec. The imposed flow is the same as that used in Figure 8.

The Stability of the Finger Doublet. Let us again address the stability of the fingers. A set of five simulations was carried out to probe this question for the finger doublet and the results are given in Fig. 8. The simulations differ only in the location of the initial porosity bump as shown. The simulations show some common features such as: 1) the fingered reaction fronts tend to split into a pair of fingers; and 2) the asymmetry allows one branch to dominate the other and the dominant one tends to stay away from the top and bottom boundary of the domain.
     Fig. 8(a) shows a symmetric state.   The initial bump is centered vertically. The dissolution front is in a perfectly symmetric doublet form.   In 8(b), the initial bump is shifted toward the lower-left corner of the domain.   The bump center on the inlet wall is located at $y_c = (5/12)y_L$ where y_L is the height of the domain. This down shift of the initial bump causes an uneven growth of two branches: the lower one being much stronger than the upper one. In 8(c) the initial bump center is at $y_c = (1/4)y_L$.   This time the upper branch is even weaker such that it only buds out periodically in a way similar to that observed in Fig. 7 except the system is in an asymmetric state.   In 8(d), the initial bump center, located at $y_c = (1/12)y_L$, is even lower than in 8(c).   Now

the upper branch is stronger than the lower branch. Finally in Fig.
8(e), the initial bump center is at the lower-left corner of the
domain. Now, the upper branch is stronger than the lower one.

Another phenomenon demonstrated by Fig. 8 is reinjection. Fig.
8(b) suggests that the upper branch of the finger will reinject into
the lower one. This is confirmed in Fig. 9 which shows a system exactly
the same as that of 8(b) except that it is longer than 8(b) by 15%.
For clarity, the first half-height contour line drawn in Fig. 9 is the
same as the last contour shown in 8(b). Also, the time interval
between each half-height contour in Fig. 9 is 2/5 of that used for
8(b). Apparently a side branch is generated but the faster advancing
main shoot creates a pressure gradient that attracts the side shoot
back into the main flow. After the upper branch is reinjected into the
lower one, the overall profile of the dissolution front in Fig. 9 is
quite similar to that for Fig. 8(c). This implies that a budding state
actually involves reinjection of a small branch back into the main
flow.

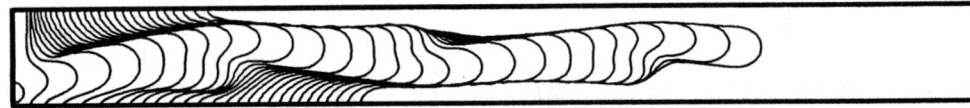

Figure 10. Meandering of the dissolution front finger. The center of
the initial porosity bump is placed on the inlet wall at $y_c = (1/6)y_L$.
The size of the domain is 60 X 6 cm, the same width as that of Figure 8
but three times as long as that of Figure 8. The time interval between
adjacent contours is 1.5 X 10^8 sec and the imposed flow is the same as
that used for Figure 8.

Fig. 10 shows a simulation for an extended domain that is two times
longer than that of 8(d) and the space resolution is reduced by a
factor of two so as to cope with computing limitation. The location of
the bump center is at $y_c=(1/6)y_L$. Under this coarser spatial gridding,
we could not resolve the lower branch of the finger as seen in 8(d). In
Fig. 10, the dissolution front meanders within the flow channel. This
meandering dissolution finger starts from a small local nonuniformity
near the inlet while the flow channel is otherwise uniform. Apparently
when the inlet flow is strong enough, a single nonuniformity may
develop into two branches of dissolution fingers. If the initial
nonuniformity is placed off-center, an oscillatory competition between
the two putative dissolution fingers may cause the dissolution front to
"meander" within the channel. In a three dimensional channel with

cylindrical symmetry, such a meandering phenomenon likely leads to a helical structure.

The above calculations demonstrate that reactive infiltration fronts support a very rich class of steady and dynamic structures. Each of these states exists in a well defined domain of system parameter space. At a given point in this space, the system may support multiple states. Which state is attained at long times depends on the state of the system in the remote past (i.e. on perturbations experienced by the front far upstream).

SUMMARY AND RELATED PROBLEMS. We have carried out numerical simulations of water-rock interaction in flow channels containing a calcite cemented sandstone. The two-dimensional reaction-transport equations have been applied to describe the interaction in which an inlet fluid undersaturated with calcite flows through the channel. It is demonstrated that a small local higher permeability nonuniformity located in the otherwise uniform channel can be strongly amplified to produce non-planar reaction fronts. The resulting front may take on half, single or multiply fingered morphologies. Competition between multiple possible fingers may cause the front to meander within the channel. The important factors that determine the ultimate reaction front profile are the speed of the inlet flow and the width of the aquifer, both of which have to be higher than critical values in order for the non-planar morphologies to develop. The initial location of the nonuniformity may strongly affect the ultimate front profile as well, especially in the cases of wide channels with large imposed flow where many basins of attraction may exist. A rich array of front morphologies can be generated by a small localized permeability nonuniformity. We believe that an even greater richness of phenomena will be observed when the simulations are done for multiple mineral or three dimensional systems. Preliminary results for multiple mineral systems are presented elsewhere.[23] Reaction-transport modeling as presented here will serve as an important guide in delineating the realm of the possible and aiding us in identifying these phenomena in natural systems or in association with enhanced oil recovery techniques such as acid flooding.

The coupling of mineral reaction to flow through Darcy's law allows for a variety of flow self-focusing phenomena that may differ from that studied here. When the fluid dissolves the matrix completely, dissolution may lead to the creation of elongate holes as observed in nature (termed karst terrains) or in association with acid stimulation as a technique to enhance petroleum recovery.[26] Furthermore, buoyancy driven flows may involve the creation of Bénard-type cells with highly focused updrafts through reaction-flow coupling. These self-localized plumes may be an important part of the dynamics of magmas rising through the earth's crust[27] or of geothermal fluid migration. Furthermore, flow self-focusing can lead to the puncturing of diagenetic or other petroleum reservoir seals when pressure differences develop across these seals; such puncturing can, when coupled to fracturing, lead to the oscillatory flow of fluids through the seal via a fracture, healing reaction cycle.[28] A number of these phenomena have been discussed further in Ref. 10. Clearly this area of research is a promising direction for the future.

References

1. J. Chadam, P. Ortoleva and A. Sen, Reactive infiltration instabilities, I.M.A. J. Appl. Math., 36 (1986) 207–221.

2. J. Chadam, P. Ortoleva and A. Sen, A weakly nonlinear stability analysis of the reactive infiltration interface, SIAM J. Appl. Math., 48 (1988) 1362–1378.

3. P. Ortoleva, E. Merino, J. Chadam and A. Sen, Geochemical self-organization II: the reactive-infiltration instability, Am. J. Sci., 287 (1987) 1008–1040.

4. G. Auchmuty, P. Ortoleva, J. Chadam, E. Merino and E. Ripley, The structure and stability of propagating redox fronts, SIAM J. Appl. Math., 46 (1986) 588–604.

5. P. Ortoleva, G. Auchmuty, J. Chadam, E. Merino, C. Moore and E. Ripley, Redox front propagation and banding modalities, Physica, 19D (1986) 334–354.

6. E. Ripley, E. Merino, C. Moore and P. Ortoleva, Mineral zoning in sediment-hosted copper deposits, in Handbook on Strata-Bound and Stratiform Ore Deposits Vol. 13, K.H. Wolf, ed., Elsevier, Amsterdam, 1985, 237–260.

7. H.K. Haskin, C.H. Moore and P. Ortoleva, Modeling acid stimulation of the Halfway Formation, Canada, using a geochemical computer model, Soc. Petr. Eng., 18133 (1988) 283–294.

8. C.H. Moore and P. Ortoleva, The effect of fluid and rock compositions on diagenesis: A modelling investigation, in Prediction of Reservoir Quality Through Chemical Modeling, I. Meshri and P. Ortoleva, eds., AAPG Memoir, 49 (1990) 103–130.

9. P. Ortoleva, E. Merino, J. Chadam and C.H. Moore, Geochemical self-organization I: Feedback mechanisms and modeling approach, Am. J. Sci., 287 (1987) 979–1007.

10. P. Ortoleva, Geochemical Self-Organization, Oxford University Press, New York, 1992.

11. J. Glimm, B. Lindquist, O. McBryan and G. Tryggvason, Sharp and diffuse fronts in oil reservoirs: Front tracking and capillarity, in SIAM, Proc. Math. and Comp. Methods in Seismic Exploration and Reservoir Modelling, Houston, January 1985.

12. J. Douglas, D. Peaceman and H. Rachford, A method for calculating multi-dimensional immiscible displacement, AIME Trans., 216 (1959) 297.

13. E. Merino, P. Ortoleva and P. Strickholm, Kinetics of metamorphic layering in anisotropically stressed rocks, Am. J. Sci., 282 (1982) 617.

14. T. Dewers and P. Ortoleva, Geochemical self-organization III: A mean field, pressure solution model of spaced cleavage and metamorphic segregational layering, Am. J. Sci., 290 (1990) 473-521.

15. T. Dewers and P. Ortoleva, The self-organization of mineralization patterns in metamorphic rocks through mechano-chemical coupling, J. Phys. Chem., 93 (1989) 2842-2848.

16. M. Gitterman and V. Steinberg, Instability of ionization equilibrium of a weakly ionized three-component plasma, Phys. Rev., A20 (1979) 1236.

17. P. Ortoleva and J. Ross, Theory of the propagation of discontinuities in kinetic systems with multiple time scales: Fronts, front multiplicity, and pulses, Chem. Phys., 63 (1975) 3398.

18. D. Feinn and P. Ortoleva, Catastrophe and propagation in chemical reactions, J. Chem. Phys., 67 (1977) 2119.

19. S. Schmidt and P. Ortoleva, Electric field effects on propagating BZ waves: Predictions for an Oregonator and new pulse supporting models, J. Chem. Phys., 74 (1981) 448.

20. R. Sultan and Ortoleva, P., Static reaction-diffusion structures in folded slow manifold systems, J. Chem. Phys., 85 (1986) 5068-5075.

21. C. E. Harvie, N. Moller and J.H. Weare, The prediction of mineral solubilities in natural waters: The $Na-K-Mg-Ca-H-Cl-SO_4-OH-HCO_3-CO_3-CO_2-H_2O$ system to high ionic strengths at $25°$ C, Geochim. Cosmochim. Acta, 48 (1984) 723-751.

22. P. C. Lichtner, The quasi-stationary state approximation to coupled mass transport and fluid-rock interaction in a porous medium, Geochim. Cosmochim. Acta, 52 (1988) 143-165.

23. W. Chen and P. Ortoleva, Reaction front fingering in carbonate cemented sandstones, in Self-Organization in Geological Systems: Proceedings of a Workshop held 26-30 June 1988, University of California Santa Barbara, P. Ortoleva, B. Hallet, A. McBirney, I. Meshri, R. Reeder and P. Williams, eds., Earth Science Reviews, 29 (1990) 183-198.

24. J. Bear, Dynamics of Fluids in Porous Media, Elsevier, Amsterdam, 1972.

25. I. Marsden and M. McCracken, The Hopf Bifurcation and Its Applications, Springer Verlag, New York, 1976.

26. M. L. Hoefner and H.S. Fogler, Pore evolution and channel formation during flow and reaction in porous media, AIChE Journal, 34 (1988) 45-54.

27. A. R. McBirney and E. L. Sonnenthal, Metasomatic replacement in the Skaergaard Intrusion: Preliminary observations, in preparation.

28. W. Chen, A. Park, A. Ghaith, and P. Ortoleva, Diagenesis through coupled processes: modeling approach, self-organization and implications for exploration, in Prediction of Reservoir Quality Through Chemical Modeling, I. Meshri and P. Ortoleva, eds., AAPG Memoir, 49 (1990) 103–130.

CHAPTER 5

Numerical Statistical Mechanical Studies of the
Behavior of Simple Fluids in Structured Slit
Micropores*

J. H. Cushman**
D. J. Diestler**
M. Schoen†

Abstract. The behavior of fluids in molecularly narrow (less than
30 Å) micropores (vicinal fluid) differs markedly from that of their
bulk-phase counterparts. Although they play an important role in a
variety of technological and natural processes, such vicinal fluids are
poorly understood. This article discusses the application of numerical
statistical mechanics to the study of vicinal fluids. Specifically, the
properties of simple fluids confined to planar micropores are examined
by means of iso-strain grand-canonical Monte Carlo and iso-strain
microcanonical molecular dynamics techniques. The equilibrium structure
of the inhomogeneous pore fluid and its influence on self-diffusion and
stress-strain relations are of principal concern. A rigorous
generalized (molecular) hydrodynamic description of the pore fluid is
developed.

1. **Introduction.** In recent years numerical statistical mechanics
has become a powerful technique for investigating the molecular behavior
of fluids [1,2]. It is especially useful for studying the behavior of
fluids in pores of width less than 100 Å, where macroscopic laboratory
experiments quite often produce ambiguous results. In this article we
review some of the recent progress our group has made on simulating the
behavior of fluids in model structured planar micropores, or slit-pores.
These models are prototypes for the more complex clay-water mixtures
found in geologic formations [3], which are of primary interest to our
group.

**Department of Agronomy and Department of Chemistry, Purdue
University, West Lafayette, IN.

† Institüt für Experimentalphysik, Universität Witten/Herdecke,
Witten, Germany.
*Support for this research was provided by grants DE-FG02-85ER-60310 of the USDOE and DAAL03-90-G-0074 of
the USARO.

The remainder of this article is organized as follows. Section 2 reviews some fundamentals of equilibrium statistical mechanics. In Section 3 the theory behind self-diffusion in a slit-pore is summarized. Using results from the theory of integral equations, we show that the memory-function formalism underlying the molecular description of diffusion is embeded in a well posed initial-value problem. We further demonstrate rigorously that the generalized diffusion tensor [4] tends toward the velocity autocorrelation function in the limit of long wavelength ($|k| \to 0$). While both of these results have been "formally" illustrated previously, we are not aware that they have been placed on a rigorous mathematical basis. Monte Carlo and molecular dynamics techniques are developed in Sections 4 and 5. Section 6 details the procedures employed and the properties computed. The validity of the canonical measure is discussed in Sec. 7 and the ergodicity of the Markov chain in Sec. 8. Section 9 presents stress-strain relationships while Section 10 summarizes diffusive behavior in the slit-pore.

   2. Statistical Mechanical Preliminaries. We begin by reviewing relevant facets of classical equilibrium statistical mechanics. The system has a time-independent Hamiltonian H defined over a phase space $(\Omega, S, P(d\omega))$, such the the expected (observed) value of any dynamical variable, assumed to be a square-integrable function on Ω, is given by

$$E(f) = (f, 1) = \int_{\Omega} f(\omega) \ P(d\omega), \tag{1}$$

where S is an appropriate σ-algebra of subsets of Ω. We assume that in the thermodynamic limit expected values over the different ensembles are equal, provided the ensembles describe the same thermodynamic state. (This assumption of "equivalence of ensembles" has not in general been rigorously justified [5]). We further assume the flow is ergodic [6], so that

$$\lim_{T \to \infty} \int_{0}^{T} f(\omega_t) \ dt/T = E(f) \tag{2}$$

where $\omega_t = T_t\omega_o$ is the flow, ω_o is the initial state in phase-space, and the limit is in the L_2-sense. This last result (Von Neumann's Ergodic Theorem) or a stronger version (Birkoff's Ergodic Theorem) is used in this article to supply some justification for the assumption of equivalance between ensemble averages and time averages.

   For simplicity we assume the system to be composed on N identical atoms of mass m. The Hamiltonian is taken to be of the standard form

$$H(\omega) = H(p_1, \ldots, p_N; x_1, \ldots, x_N)$$

$$= \sum_{i=1}^{N} p_i^2/2m + U(x_1, \ldots, x_N), \tag{3}$$

where $\{x_i\}_{i=1}^N$ is the set of position coordinates (configuration) and $\{p_i\}_{i=1}^N$ is the set of conjugate momentum coordinates. We define a phase space Ω_N by

$$\Omega_N = R^{3N} \times V^N \tag{4}$$

where V (assumed measurable) is the volume of the system and R^{3N} refers to the momentum subspace. Throughout this article the potential energy U is taken to be a sum of pairwise Lennard-Jones (12,6) interactions,

$$U = \sum_{i=1}^{N-1} \sum_{j>i} u(x_{ij})$$

$$u(x) = 4\varepsilon[(\sigma/x)^{12} - (\sigma/x)^6] \tag{5}$$

where $x_{ij} = |x_i - x_j|$, ε is the depth of the attractive well and σ is approximately the hard-sphere diameter of an atom.

The Gibbs ensembles of primary interest to us are the grand-canonical and microcanonical. We mention the canonical ensemble in passing only because of its involvement in the generation of a realization of a Markov chain of configurations that are distributed according to the probability measure for the grand ensemble. Let

$$\lambda_N^c(d\omega) = e^{-\beta H(\omega)} d\omega/N! h^{3N} \tag{6}$$

and

$$\lambda_N^m(d\omega) = \delta(H(\omega) - E) d\omega/N! h^{3N}, \tag{7}$$

where E is a fixed energy, $\beta = 1/kT$, k is the Boltzmann constant, δ is the Dirac measure and h is Planck's constant. The probability measure for the microcanonical ensemble is given by

$$P_N^m(d\omega) = \lambda_N^m(d\omega)/\lambda_N^m(\Omega_N) \tag{8}$$

and that for the canoncial ensemble is

$$P_N^c(d\omega) = \lambda_N^c(d\omega)/\lambda_N^c(\Omega_N). \tag{9}$$

The grand canonical probability measure P^g is defined over $\bigoplus_{N=0}^{\infty} \Omega_N$. The restriction of P^g to Ω_N is given by

$$P_N^g(d\omega) = e^{N\beta\mu} \lambda_N^c(d\omega)/ \sum_{I=0}^{\infty} e^{I\beta\mu} \lambda_I^c (\Omega_I) \tag{10}$$

where μ is the chemical potential, $R^o \times V^o$ is a point and $\lambda_0^c (\Omega_0) = 1$.

The reader should note that expression (10) is well defined because of the *stability* of the Lennard-Jones potential [5].

The expected value of any dynamical variable not depending on momenta can be computed over configuration space alone. We accomplish this by explicitly integrating over the momentum coordinates and leaving only the configurational integrals. We may thus define a (reduced) probability measure over configuration space for the canonical ensemble by

$$\lambda_N^{c'}(dx) = e^{-\beta U(x)} \, dx_1 \, \ldots \, dx_N/N! \tag{11}$$

$$P_N^{c'}(dx) = \lambda_N^{c'}(dx)/\lambda_N^{c'}(v^N) \tag{12}$$

The configurational grand-canonical probability space is defined by a probability measure $p^{g'}$ over $\overset{\infty}{\underset{N=0}{\oplus}} v^N$. The restriction to v^N of $p^{g'}$ is given by

$$P_N^{g'}(dx) = a^N \, v^N \, \lambda_N^{c'}(dx)/\sum_{I=0}^{\infty} v^I \, a^I \lambda^{c'}(v^I) \tag{13}$$

where the activity a is

$$a = e^{\beta \mu} \, (2\pi m/h^2 \beta)^{3/2} \tag{14}$$

We define the configurational partition function by

$$Z(V,N,\beta) = \lambda_N^{c'}(v^N) \tag{15a}$$

and the grand-canonical partition function by

$$\Xi(V,a,\beta) = \sum_{N=0}^{\infty} v^N \, a^N \lambda_N^{c'}(v^N) \tag{15b}$$

Henceforth, the primes are dropped when no confusion can arise. From (15) it follows that

$$\Xi(V,a,\beta) = \sum_{N=0}^{\infty} a^N Z(V,N,\beta) \tag{16}$$

where $Z(V,0,\beta) = 1$.

3. **Self-Diffusion.** Several very interesting questions concerning the diffusion of fluid atoms in slit-pores naturally arise:

   i)   Does a diffusing atom obey Fick's first and second laws?
   ii) How wide must a pore be for the continuum approximation to hold?
   iii) To what extent is the diffusion tensor anisotropic?
   iv) How are the components of the diffusion tensor related to the pore wall structure?

v) Is the diffusion tensor affected by the alignment of one
    pore wall with respect to the other?

All of these questions have been addressed in the recent literature [7].
In this section we present the basic machinery needed to tackle the
first and second questions and defer consideration of the remaining
questions until section 10.

If the time and space scales are suitably large so that Fick's
first and second laws hold for a diffusing atom in a bulk fluid and if
the conditional probability density, G, of finding an atom in $d\mathbf{x}$ at
$(\mathbf{x},t)$, given that it is initially at $\mathbf{x}_0$, is C^2 with finite Fourier
transform over R^3, then it is well known [8] that as t gets large

$$<(\Delta x)^2> = <(\Delta y)^2> = <(\Delta z)^2> \rightarrow 2Dt \qquad (17)$$

where $\Delta x = x - x_0$ is the displacement in the tagged atom's x-coordinate.
The conditional probability density is of course given by the normal
distribution. If, however, Fick's first and second laws hold for a
fluid between two infinite plane-parallel walls, aligned normal to the
z-coordinate, it is not necessarily true that

$$<(\Delta z)^2> \rightarrow 2Dt \qquad (18)$$

in the limit of large t. We have, however, the following result.
Theorem 1: If $c(z,t)$ satisfies the one-dimensional diffusion
equation with a point source at $(z_0, 0)$ and no-flux boundary conditions
at z = 0 and z = h, then the conditional probability density, G, for a
diffusing atom initially at the origin, may be constructed from $c(z,t)$.
Moreover,

i) $$G(z',t) = (h \pm z')/h^2 + \sum_{j=1}^{\infty} \exp(-j^2\pi^2 D_{zz}t/h^2)$$

$$\times \left[\frac{h \pm z'}{h^2} \cos (j\pi z'/h) \pm \frac{1}{j\pi h} \sin(j\pi z'/h) \right] \qquad (19)$$

ii) $$<(\Delta z)^2(t)> = h^2/6 - 16h^2/\pi^4 \sum_{j=1}^{\infty} (2j - 1)^{-4} \exp[-(2j-1)^2\pi^2 D_{zz}t/h^2] \qquad (20)$$

iii) $$\lim_{h\to\infty} <(\Delta z)^2(t)> = 2D_{zz}t \qquad (21)$$

iv) $$<(\Delta z)^2(t)> = O(h^2) \text{ as } t \to \infty \qquad (22)$$

where the +(-) sign in (19) holds for z' > 0 (z' < 0), $z' = z - z_0$, and
D_{zz} is the z-component of the assumed diagonal diffusion tensor.

Pf.
Part (i) follows by solving the one-dimensional diffusion equation
for $c(z, t; z_0)$, translating c to the origin and defining G as the
average of c over all possible origins [9].
Part (ii) follows from (i) by multiplying G by z'^2, integrating on
z' from -h to h and invoking the uniform convergence of the series to

justify term by term integration. Parts (iii) and (iv) also follow from the uniform convergence of G as t → ∞ and h → ∞, respectively.

QED

While the theorem allows us to estimate how wide the pore must be in order for diffusion perpendicular to the walls to become essentially free (as in the bulk), it cannot tell us how wide the pore needs to be for the diffusion to be Fickian. The following development provides the means to do this. Our first task is to set up a well posed intial-value problem to define a general wave-vector and frequency-dependent diffusion tensor valid on any space-time scale. To do so we first recall the following result [10].

Let B be a convolution operator of the form

$$Bw = \int_0^t K(\tau) \ w(t-\tau) d\tau \tag{23}$$

and assume $w \varepsilon C^1[0,\infty)$. Then if $K \varepsilon L_1(Loc)$ the problem

$$\dot{w} = Bw, \tag{24}$$

$$w(0) = w_0 \tag{25}$$

has a unique solution on $0 \leq t < \infty$.

It can be shown [1] that the correlation function, ψ, of any dynamical variable, u, over phase space (i.e., any square-integrable function over $R^{3N} \times V^N$) satisfies a Volterra equation of the form

$$\dot{\psi} = - \int_0^t K(\tau) \ \psi(t - \tau) \ d\tau \tag{26}$$

with $\psi(0) = 1$. The kernel, or memory function, takes the form

$$K(\tau) = (\dot{u}, \ e^{i\tau QL} \ \dot{u}) \tag{27}$$

Here L is Liouville's operator [1] and Q is a projection operator defined by

$$Q = I - P \tag{28}$$

where the action of P on f is given by

$$Pf = u(u,f), \ \forall \ f \in L_2(\Omega,P). \tag{29}$$

Now from (27) it follows immediately that

$$|K(\tau)| \leq \|\dot{u}\|^2 \tag{30}$$

Hence from our previous discussion we have

Theorem 2: If $u \ \varepsilon \ H_2(\Omega,P)$, $\|\dot{u}(t)\| \in L_1(Loc)$,

$$K(t) = (\dot{u}, \ e^{itQL} \ \dot{u}) \tag{31}$$

where Q and L are as previously defined, and if

$$\psi(t) = (u(0), \ u(t))/\|u(0)\|^2 \in C^1[0,\infty) \tag{32}$$

then $\psi(\tau)$ is the unique solution of

$$\dot{\psi} = - \int_0^t K(\tau) \ \psi(t - \tau) \ d\tau \tag{33}$$

$$\psi(0) = 1 \tag{34}$$

Practical algorithms for the solution of (33) may be found in Berne and Harp [11].

Zwanzig [4] has shown that if

$$u = e^{ik \ \cdot \ x(t)} \tag{35}$$

then $K_k(t)$ as defined by (31) satisfies

$$K_k(t) = k \ \cdot \ D(k,t) \ \cdot \ k \tag{36}$$

Here k is a vector in reciprocal space and $D(k,t)$ is a generalized diffusion tensor obeying a generalized Fick's first law

$$\tilde{j} = \tilde{D}(k,\omega)(\nabla c)_{k,\omega} \ , \tag{37}$$

where ~ denotes the Fourier-Laplace transform and j is the flux of c. Moreover, Zwanzig [3] has shown that

$$D(k,t) = ([v(0) \ e^{-ik \ \cdot \ x(0)}], \ e^{it \ Q_k L} \ [e^{ik \ \cdot \ x(0)} \ v(0)]), \tag{38}$$

where $v = \dot{x}$.

Theorem 3: If $D(k,t)$ is as defined above, then

$$\lim_{|k|\downarrow 0} D(k,t) = (v(0), \ v(t)), \tag{39}$$

that is, at small wavevectors the generalized diffusion tensor tends to the velocity autocorrelation function.

Pf.

We want to look at

$$\lim_{|k|\downarrow 0} (v(0) \ e^{-ik\cdot x(0)}, \ e^{itQ_k L} \ [e^{ik\cdot x(0)} \ v(0)]).$$

Clearly

$$\|e^{itQ_kL}\| = 1, \ \forall \ k$$

and hence

$$|(v(0) \ e^{-ik \cdot x(0)}, \ e^{itQ_kL} \ [e^{ik \cdot x(0)} \ v(0)])|$$

$$= |(v(0) \ \sum_{j=0}^{\infty} \frac{(-ik \cdot x(0))^j}{j!}, \ e^{itQ_kL} \ \left\lceil \sum_{\ell=0}^{\infty} \frac{(ik \cdot x(0))^{\ell}}{\ell!} \ v(0) \right\rceil)|$$

$$= |(v(0), \ e^{itQ_kL} \ v(0))| + 0(|k|) \tag{40}$$

Next note that by the dominated convergence theorem, P_kf is a continuous function of k at $|k| = 0$, and consequently

$$\lim_{|k| \downarrow 0} P_kf = P_0f = E(f), \ \forall \ f \in L_2(\Omega, P) \tag{41}$$

where $E(f)$ is the expected value of f. Let $|k|^{-1} \in \{1, 2, \ldots\}$. Then from (41) and the "approximation theorem" for semigroups [12] it follows that

$$\lim_{|k| \downarrow 0} e^{itQ_kL} \ f = e^{it(I-E)L} \ f, \ \forall \ f \in L_2(\Omega, P) \tag{42}$$

We conclude from (40) and (42) and the dominated convergence theorem that

$$\lim_{|k| \downarrow 0} D(k,t) = (v(0), \ e^{it(I-E)L} \ v(0)) \tag{43}$$

Now we have

$$e^{it(I-E)L} \ v(0) = \sum_{j=0}^{\infty} \frac{[it(I-E)L]^j}{j!} \ v(0)$$

$$= v(0) + it(I-E)\dot{v}(0) + \frac{(it)^2}{2!} (I-E)L(I-E) \ \dot{v}(0) + \ldots$$

$$= v(0) + it \ \dot{v}(0) + \frac{(it)^2}{2!} (I-E) \ \ddot{v}(0) + \ldots$$

$$= v(0) + it \ \dot{v}(0) + \frac{(it)^2}{2!} \ \ddot{v}(0) + \frac{(it)^3}{3!} \ \dot{\ddot{v}}(0) + \ldots$$

$$= e^{itL} \ v(0) \tag{44}$$

where we have used the identities $\dot{f} = Lf$ and the fact that the expected value is over the equilibrium ensemble. But

$$e^{itL} \mathbf{v}(0) = \mathbf{v}(t) \tag{45}$$

and hence

$$\lim_{|\mathbf{k}| \downarrow 0} D(\mathbf{k}, t) = (\mathbf{v}(0), \mathbf{v}(t)) \tag{46}$$

QED.

We note that a slightly stronger formal result indicates the convergence in the last theorem is $O(k^2)$ [7].

In addition to the classical result (17), one also has

$$D_{zz} = \int_0^\infty (v_z(0), v_z(\tau))d\tau, \tag{47}$$

provided that $(v_z(0), v_z(\tau))$ goes to zero faster than τ^{-2}, $v \in L_2(R^{3N} \times V^N, P)$ and V is unbounded. Thus Theorem 3 in conjunction with (47) can be used to examine the validity of the continuum approximation. That is, if

$$\int_0^\infty D_{zz}(\mathbf{k}, t)dt \approx D_{zz} \tag{48}$$

holds for small $|\mathbf{k}|$, then the continuum approximation is valid. In (48) D_{zz} on the right hand side is identical to D_{zz} appearing in (47) and on the left hand side that of (36). Alternatively we may check whether the approximation

$$\tilde{D}_{zz}(\mathbf{k}, \omega) \approx \tilde{C}_z(\omega) \tag{49}$$

holds when $|\mathbf{k}| \downarrow 0$, $|\omega| \downarrow 0$, where $\tilde{C}_z(\omega)$ is the causal Fourier time transformation (i.e. the spectral density) of the z-component of the velocity autocorrelation tensor.

4. **Monte Carlo Techniques.** The central objective in numerical equilibrium statistical mechanics is to approximate integrals of the form

$$<f> = \int_\Omega f(\mathbf{x}) P(d\mathbf{x}) \tag{50}$$

where f is an S-measurable function and P is the probability measure corresponding to the particular ensemble under investigation. Because N is generally large (> 100) the classical evaluation of (50) via deterministic numerical quadrature is impractical, if not impossible, using existing hardware. Researchers have thus turned to stochastic methods, specifically Monte Carlo techniques, to evaluate (50). In this section we briefly review the classical Metropolis [13] canonical and grand-canonical methods. To evaluate (50) efficiently, we must sample from a distribution of configurations according to their probability of occurrence. The sampling is accomplished for the canonical ensemble by generating a realization of a Markov chain of configurations of atoms, which has as its limiting distribution $P_N^c(d\mathbf{x})$, according to (12). Let $\mathbf{x}^i$ be a point in configuration space and let

$$\rho_i = e^{-\beta U(\mathbf{x}^i)}/N! \; Z(V,N,\beta) \tag{51}$$

The set of elements $\{\rho_i\}$ is the limiting distribution of the Markov chain for the canonical ensemble and the task is to construct a transition matrix Π with limiting distribution given by the frequency of occurrence of ρ_i, that is the task is to find the finite matrix Π such that

$$\rho = \rho\Pi \quad (\text{i.e. } \rho_j = \sum_{i=1}^{\infty} \rho_i \, \Pi_{ij}) \tag{52}$$

$$\sum_{j=1}^{\infty} \Pi_{ij} = 1 \quad (\text{i.e. } \Pi_{ii} = 1 - \sum_{j \neq i}^{\infty} \Pi_{ij}) \tag{53}$$

and

$$\Pi_{ij} \geq 0 \quad \forall \; i,j \; . \tag{54}$$

The standard route for constructing Π is to replace (52) with the more restrictive equation

$$\rho_j \, \Pi_{ji} = \rho_i \, \Pi_{ij} , \tag{55}$$

which expresses "microscopic reversibility". Clearly (55) implies (52). Next let α be an infinite, symmetric matrix of non-negative constants. In the Metropolis algorithm we set

$$\Pi_{ij} = \alpha_{ij} \min (1, \rho_j/\rho_i) \qquad i \neq j \tag{56}$$

$$\Pi_{ii} = 1 - \sum_{j \neq i} \Pi_{ij} \tag{57}$$

We note that α is somewhat arbitrary. Configurations are generated as follows. Atoms are placed initially in a random configuration in the simulation cell. At each successive step in the Markov chain an atom (k) is chosen at random from a uniform distribution and moved uniformly randomly in a cube R of side $2\delta R$ centered on this atom. Because of the computer's finite architecture, there are only M possible moves for atom k located at $x_k^{(i)}$ (i being the current state, or configuration), that is there are only M possible neighboring states j. Hence, α is chosen such that

$$\alpha_{ij} = 1/M \qquad x_k^j \in R \tag{58}$$

$$\alpha_{ij} = 0 \qquad\qquad x_k{}^j \notin R \qquad\qquad (59)$$

where $x_k{}^j$ is the coordinate of atom k in the neighboring state j. The elements of Π (as defined by (56)) depend on $\delta U_{ji} = U_j - U_i$ where $U_i = U(x^i)$ is the configurational energy in state i, that is

$$\Pi_{ij} = \alpha_{ij} \min (1, e^{-\beta\delta U}ji) \qquad i \neq j \qquad\qquad (60)$$

A realization of the Markov chain is constructed as follows. If $\delta U_{ji} \leq 0$, then the probability of state j is greater than that of state i and the new configuration is accepted. On the other hand, if $\delta U_{ji} > 0$ then the move is accepted with probability $\exp(-\beta\delta U_{ji})$. If the move is rejected, then the old configuration becomes the new configuration and it is counted again in accordance with (57). Finally (50) is evaluated via an ergodic assumption by averaging f over the realized configurations.

The Metropolis algorithm is not readily vectorizable, but it is highly parallelizable. Parallelization follows from generation of L independent Markov chains (one per processor), all converging to $P^c{}_N(dx)$ and using

$$\rho^{(\ell)} = \rho^{(\ell)} \Pi^{(\ell)} \qquad\qquad (61)$$

$$\rho^{(\ell)} = \rho^{(\ell)} \sum_{i=1}^{L} \Pi^{(i)}/L \qquad\qquad (62)$$

and

$$<f> = \sum_{\ell=1}^{L} <f>^{\ell}/L \qquad\qquad (63)$$

where $<f>^{(\ell)}$ is the average over configurations generated by the ℓ-th processor.

The extension of the Metropolis algorithm to the grand-canonical ensemble (GCEMC) is straightforward [14,15]. The main feature distinguishing the grand-canonical from the canonical ensemble is that the chemical potential, μ (or activity, a), is held fixed rather than the number of atoms, N. In the grand-canonical ensemble Metropolis procedure atoms are moved according to the canonical Metropolis method. However, since N is variable, we must also create and destroy atoms according to ratio ρ_j/ρ_i where

$$\rho_j = \exp[-\beta U_j - \ln (N \pm 1)! + (N \pm 1)\ln V + (N \pm 1) \ln a]/\Xi \qquad (64)$$

and

$$\rho_i = \exp[-\beta U_i - \ln N! + N \ln V + N \ln a]/\Xi \qquad\qquad (65)$$

Hence atoms are created with probability $\min (1, e^{-C}ji)$ and destroyed with probability $\min (1, e^{-D}ji)$, where

$$C_{ji} = \beta \delta U_{ji} - \ln \left[\frac{aV}{N+1} \right] \tag{66}$$

and

$$D_{ji} = \beta \delta U_{ji} - \ln \left[\frac{N}{aV} \right] \tag{67}$$

This algorithm may be parallelized as in the case of Metropolis method.

5. Molecular Dynamics and Vectorization. The microcanonical ensemble conserves energy, volume and number of atoms. For a fixed number of atoms in a fixed volume Newton's classical equations of motion also conserve energy. Consequently, as partly justified by the ergodic hypothesis stated earlier, we may equate time averages over the trajectory determined through Newton's equations with microcanonical ensemble averages. Thus, the essence of the molecular dynamics (MD) method is the evaluation of

$$<f> = \int_{\Omega_N} f(\mathbf{x}) \ P^m(d\mathbf{x}) = \lim_{T \to \infty} \frac{1}{T} \int_0^T f(\mathbf{x}_t) dt \tag{68}$$

by integrating Newton's equations and averaging relevant dynamical variables over time. These equations are

$$F_i = m \ d^2 \mathbf{x}_i / dt^2 = - \nabla_i U \qquad i = 1, \ldots, N \tag{69}$$

where F_i is the force on the i-th atom and U is given by (5). Equations (69) represent a system of 3N equations in 3N unknowns (the $\mathbf{x}_i$). There are many schemes for integrating (69); we have chosen the Störmer-Verlet [16] algorithm, which we have optimally vectorized for the CYBER 205 [17].

6. Details of Simulations and Properties. The prototypical slit-pore [18] consists of a Lennard-Jones fluid confined between two parallel planes (walls) of rigidly fixed atoms arranged in configurations identical to the (100) plane of the fcc lattice (Fig. 1). The length of the side of the wall is s. The wall-fluid interactions are identical with the fluid-fluid interactions. A laboratory coordinate frame is set up such that the walls are perpendicular to the z-axis; wall 1 corresponds to $z = z_1 = 0$. The coordinates of an atom in wall 2 are related to those in wall 1 via

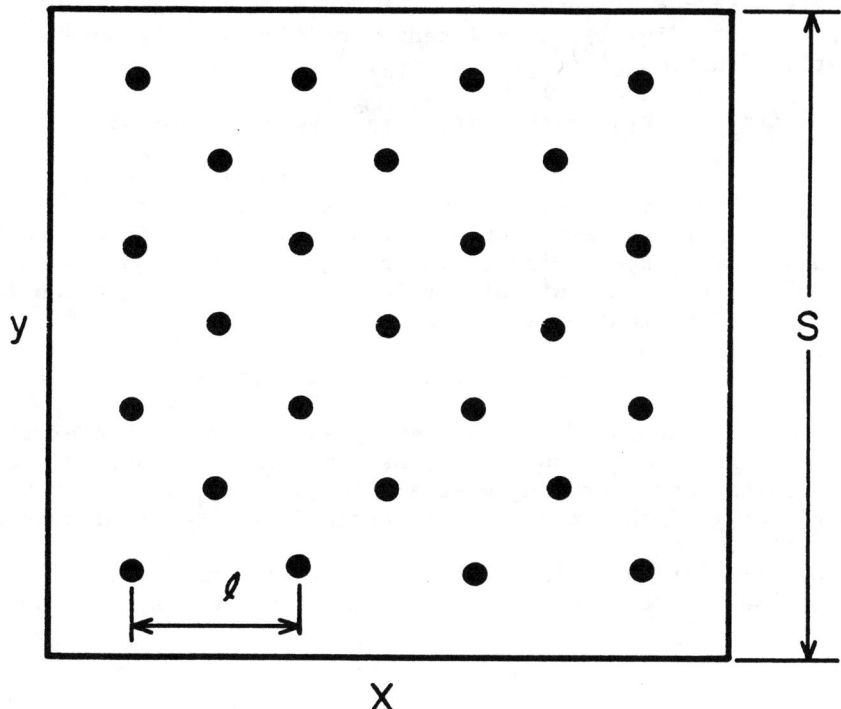

Figure 1.   A square "unit cell" of the (100) plane of the fcc lattice.

$$x_2 = x_1 + \alpha \, \ell \qquad\qquad 0 \leq \alpha < 1 \qquad\qquad (70a)$$

$$y_2 = y_1 \qquad\qquad (70b)$$

$$z_2 = z_1 + h \qquad\qquad (70c)$$

where ℓ is defined in Fig. 1.   The parameter α defines the registry of the walls.   When $\alpha = 0$, the walls are said to be in registry; when $\alpha = 0.5$, the walls are said to be completely out of registry.   By fixing α we are effectively fixing the strain on the system.   Hence we are modeling systems in an iso-strain ensemble.

The total potential energy of the system is

$$U = \sum_{i=1}^{N-1} \sum_{j>i}^{N} u(r_{ij}) + \sum_{i=1}^{N} \sum_{j=1}^{2N_s} u(r_{ij}) \qquad (71)$$

where N_s is the number of atoms in a wall.

The presence of the pore walls creates inhomogeneity and anisotropy in the pore fluid.   Consequently, the number density $\rho^{(1)}$ depends on the position x, although the symmetry of the pore walls requires $\rho^{(1)}$ to be periodic in any x-y plane.   Because of statistical error, however, it is impractical to compute the x-y dependence of $\rho^{(1)}$ by either GCEMC or MD. Consequently we take $\rho^{(1)}$ to be a function of only z.   Thus

$$\rho^{(1)}(z) = \langle N(z)\rangle/A\Delta z \qquad (72)$$

where $\langle N(z) \rangle$ is the expected number of atoms in a layer of thickness Δz at z and $A = s^2$ (Fig. 1). Consistent with (72) we introduce a pair correlation function $g^{(2)}(z_1, \rho_{12}, z_{12})$ defined by

$$g^{(2)}(z_1, \rho_{12}, z_{12}) = \langle N(z_1, \rho_{12}, z_{12}) \rangle / [2\pi \rho_{12} \Delta \rho_{12} \Delta z_{12} \rho^{(1)}(z_2)] \qquad (73)$$

Here z_1, ρ_{12} and z_{12} are respectively the z coordinate of reference atom 1, and the cylindrical coordinates of atom 2 relative to 1. $\langle N(z_1, \rho_{12}, z_{12}) \rangle$ is the expected number of fluid atoms in an annulus centered at (x_1, y_1, z_2) having radius ρ_{12}, width $\Delta \rho_{12}$ and thickness Δz_{12}. Throughout the remainder of this article we restrict attention to the in-plane pair distribution function

$$g^{(2)}(z_1, \rho_{12}, 0) = g^{(2)}(z_1, \rho_{12}) \qquad (74)$$

$\langle N(z_1, \rho_{12}, 0) \rangle$ is computed by considering each atom in a layer of the fluid of thickness Δz_1 about $z = z_1$ as a reference, counting the atoms in its annulus, and averaging over all atoms in a layer about $z = z_1$. The height Δz_{12} of the annulus is determined by $\rho^{(1)}$ as discussed in a later section.

Other properties of interest in later sections are the normal and shear stresses exerted on the walls by the fluid phase. The shear stress is given by

$$\tau_{zx} = F_x/s^2 = \sum_{i=1}^{N} \sum_{j=1}^{N_s} \langle x_{ij} \, r_{ij}^{-1} \, du/dr_{ij} \rangle / s^2 \qquad (75)$$

where $x_{ij} = x_i - x_j$. Similarly the normal stress on the wall is given by

$$\tau_{zz} = F_z/s^2 = \sum_{i=1}^{N} \sum_{j=1}^{N_s} \langle (z_{ij} \, r_{ij}^{-1} \, du/dr_{ij} \rangle / s^2 \qquad (76)$$

The normal stress is often called the "solvation force" and is related to the swelling pressure P_s by

$$P_s = \tau_{zz} - P_a \qquad (77)$$

where P_a is the pressure of bulk fluid that is in thermodynamic equilibrium with the pore fluid.

Finally, because the pore walls are supposed infinite, we invoke periodic boundary conditions in the x- and y-directions [18] and employ the minimum-image convention to compute interatomic distances. The Lennard-Jones potential is short-ranged so we truncate interactions between atoms sufficiently far apart. On account of the symmetry of the pore, we center an imaginary cutoff cylinder of radius ρ_c on each atom i and include explicitly only interactions with atoms j inside the cylinder. Contributions to averaged thermodynamic quantities from interactions between atoms separated in the x-y plane by distances greater than ρ_c are evaluated analytically under the approximation that the pair-correlation has become unity.

In the destruction-creation step of the GCEMC method the density of the fluid in the trial state (where the number of particles has increased or decreased by one) differs from that in the original state by $\pm$ 1/V. Therefore, the truncation of interactions also necessitates corrections of the computed change in the configurational energy accompanying the creation-destruction step. Expressions for these corrections are given by Schoen et al. [18].

Unless otherwise noted, throughout the remainder of the article we will employ the reduced dimensionless variables defined in Table I.

7. Validity of the Canonical Probability Measure. On the basis of their simulation of a Lennard-Jones fluid in which equipartition of energy [19] is violated, Thurtell and Thurtell [20] (hereafter denoted TT) have hypothesized that the canonical probability measure (Maxwell-Boltzmann distribution) is inappropriate for fluids in pores. We have argued [21] their observed violation of the equipartition theorem results from a defect in their implementation of a diffuse reflection condition at the pore walls: we have shown that, if the diffuse reflection condition is implemented properly, the MD microcanonical method obeys the equipartition theorem. This does not in itself prove the applicability of the canonical probability measure, which, however, can easily be done for any classical, conservative ergodic system (such as TT's), as shown in standard textbooks [19].

TT have also argued that it is inapproprate to integrate away the momentum coordinates to arrive at the configurational probability measure when computing expected values of quantities depending only on configurational coordinates. But, as demonstrated in Sec. 2, this is clearly false. Indeed, this result is the basis of the Monte Carlo methods presented. Moreover, we have shown [7] that GCEMC and MD agree when systems are in the same thermodynamic state (i.e., the equivalence of ensembles assumption of Section 2 is valid).

8. Ergodicity of the Markov Chain. The subject of phase transitions in micropores is currently under intensive investigation. Experimental sorption isotherms exhibit hysteresis, that is, the adsorption branch of the isotherm does not coincide with the desorption branch. Hysteresis is commonly interpreted using the hydrodynamic model of Saam and Cole [22], which relates the thermodynamic state of the pore fluid to the behavior of the long wavelength modes as a function of the thickness of the adsorbed layer. Sorption isotherms have also been studied by numerical equilibrium statistical-mechanical [23,24,25] methods. However, doubt has recently been cast by the authors [26] on the validity of such methods in the study of hysteresis. This doubt owes its origin, as we now illustrate, to the ergodic behavior (or rather lack of it) of the Markov chain near critical points of the phase diagrams.

Hysteresis in the slit-pore model can be represented as a cusp catastrophe. Figure 2 displays a qualitative plot of n^* as a function of μ^* and h^* (which appears as a surface in three-dimensional space) for fixed T^* and s^*. The projection of this surface onto the μ^* - h^* plane shows the bifurcation set, which corresponds to the hysteretic region in the vicinity of the capillary-condensation transition. The bifurcation set represents metastable thermodynamic states of the pore fluid, the vertex corresponding to the capillary critical point. The path EF above the critical point is reversible; n^* is a continuous function of μ^* along EF. The isotherms (calculated using GCEMC [26]) in Fig. 3 for h^*

CUSP CATASTROPHE

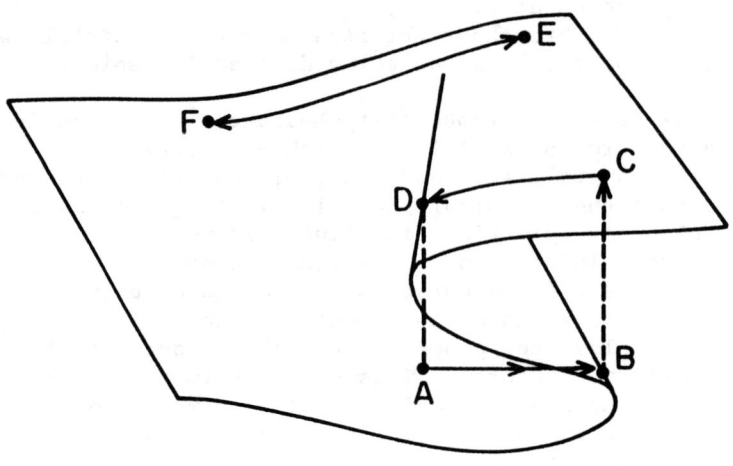

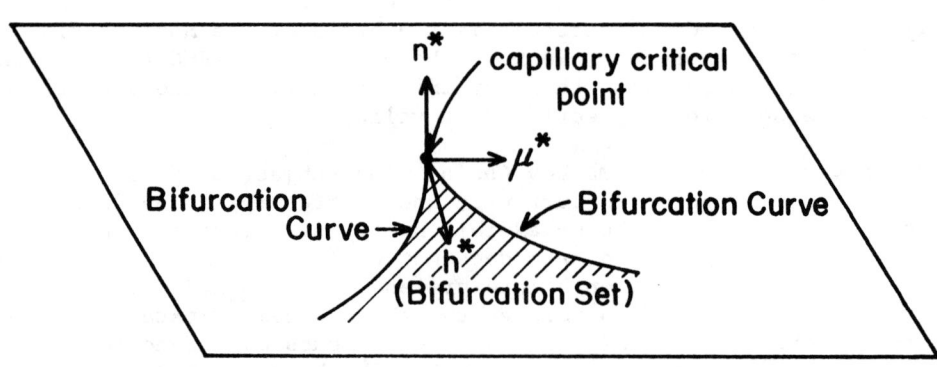

Figure 2.   The cusp catastrophe representing hysteresis in the slit-pore
model.

= 2.15 and 3.05 exemplify such a reversible path. On the other hand, if
the path traverses the bifurcation set, it is irreversible. To see
this, suppose the pore fluid is initially liquid in thermodynamic state
C (Fig. 2). As μ^* is gradually decreased, the state of the system moves
along CD. At D the liquid abruptly (i.e. catastrophically) evaporates
to form gas in state A. This liquid-to-gas transition corresponds to
the left bifurcation curve in the μ^* - h^* plane. If μ^* is next
increased, the gas in state A does not instantaneously condense, but
rather persists along AB until state B is reached, whereupon it suddenly
condenses to liquid in state C. The gas-to-liquid transition is
represented by the right bifurcation curve in the μ^* - h^* plane. The
isotherm for h^* = 4.9, s^* = 7.9925 plotted in Fig. 3 is a realization of
the type of irreversible path just described. The loop CDABC in Fig. 2
is the counterpart of the loop in Fig. 3.

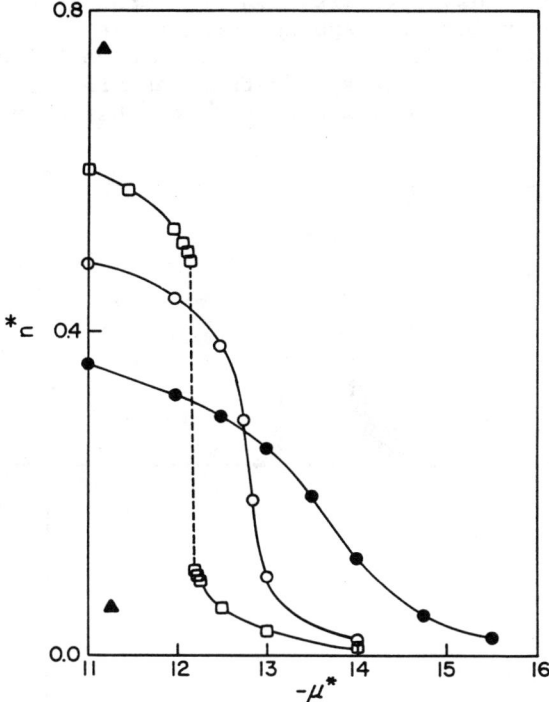

Figure 3.   Sorption isotherms for a pore at T^* = 1.0.   (a) h^* = 2.155
($\bullet$ $\bullet$ $\bullet$); (b) h^* = 3.05 (o o o); (c) h^* = 4.9 ($\square$ $\square$ $\square$).   Dashed line
indicates capillary-condensation transition for h^* = 4.9.   Two points
(Δ) on the sorption isotherm of bulk fluid are shown.   From Schoen et
al. [23].

     Unfortunately the results presented in Fig. 3 are highly dependent
on the length of Markov chain and on s^* [26], the reduced side length of
the computational unit cell (Fig. 1).   If s^* is increased, as well the
length of the Markov chain, then the bifurcation set depicted in Fig. 2
is reduced in size and in the limit as $s^* \to \infty$ this set should degenerate
to a straight line reflecting no hysteresis.
     Calculations for the slit-pore at T^* = 0.7 and h^* = 4.9 reveal a
liquid-solid transition in the range $-10.4 > \mu^* > -11.0$.   Also at $T^* =$
0.7 and h^* = 4.9 there is a gas-solid transition in the range $-9.5 > \mu^*$
> -12.1.   The sorption isotherms are strongly hysteretic in these
ranges, more so near the gas-solid than the liquid-solid transition.
These results are explained via Fig. 4, which displays plots of the
normalized fraction of Monte Carlo configurations as a function of
density.   Note that the distribution for the solid phase has zero width
(Fig. 4b).   The system was started with a solid-like configuration of
360 atoms and all accepted configurations differed by not a single atom
in 4 million steps.   The density fluctuations are extremely small and
hence an astronomical number of steps would be required to broaden the
distribution sufficiently to overlap with the tails of the liquid- or
gas-phase distributions.   The overlapping of the distributions implies
the possibility of the system's undergoing a phase transition.   Figure
4a shows that the liquid- and gas-phase distributions nearly overlap one
another in the region of intermediate density.   Hence, if the system

were started in the gas phase, it might undergo transition to the liquid phase in a reasonable number of steps and vice versa. It appears that hysteresis occurs when the density of the system cannot fluctuate sufficiently, which precludes the system from assuming with the proper frequency all configurations consistent with its thermodynamic state.

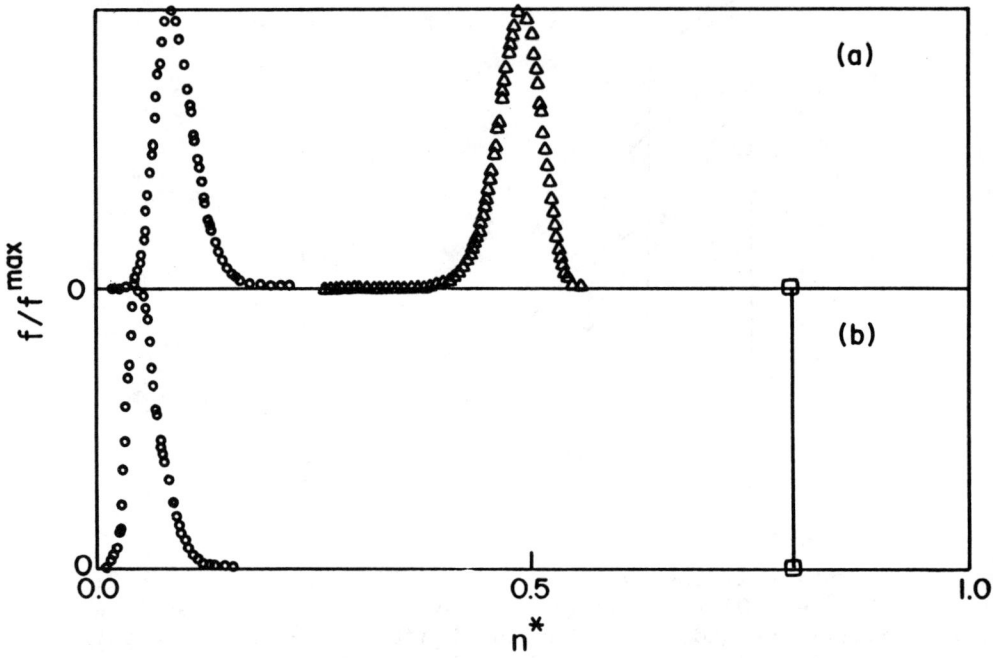

Figure 4. Normalized fraction f/f^{max} of Monte Carlo configurations as a function of the density n^*. (a) $h^* = 4.9$, $T^* = 1.0$, $s^* = 9.591$, $\mu^* = -12.225$ (O O O) and $\mu^* = -12.175$ (Δ Δ Δ); (b) $h^* = 4.9$, $T^* = 0.7$, $s^* = 9.591$, $\mu^* = -9.52$ (O O O) and $\mu^* = -9.5$ ($\square$ $\square$ $\square$). From Schoen et al. [23].

In summary the Markov chain is non-ergodic (reducible) over the number of configurations generated. Recall that a Markov chain is ergodic if any state having non-zero probability of occurrence can be reached in a finite number of steps from any other state having a non-zero probability of occurrence [27]. In principle, the GCEMC Markov chain is irreducible. In practice, however, the system may get trapped in a localized region of configuration space. In the number of steps available, dictated by one's computer budget, it may not be able to reach other regions that have significant probabilities of occurrence. Fortunately, it is our experience that this problem is limited to regions of the phase diagram near critical points.

9. **Stress-Strain Relationships.** Our interest in this section is in the normal component of strain, governed by h, and the shear component of strain, determined by α, in relation to the solvation force (normal stress) τ_{zz} and the shear stress τ_{zx}, respectively.

Figure 5 shows a plot of τ^*_{zz} vs h^* for the data of Table II as computed with GCEMC. For this particular figure, the system was equilibrated for approximately 10^6 configurations before averages were accumulated over the following $4 \cdot 10^6 - 10^7$ configurations. The cutoff radius is $\rho_c = 3.5\ \sigma$. The force oscillates as a function of h^* with a period of approximately one particle diameter. This oscillation persists out to approximately six diameters; thereafter the curve monotomically decreases. There is a direct correlation between minima in Fig. 5 and large "discontinuities" in the n^* and h^* curve. If the fluid were uniformly layered, the number of atoms leaving the pore at each of these jumps would be approximately the same as the number of atoms in a layer. Between these jumps the number density is approximately constant. In other words, the number of fluid atoms residing within the pore is quantized. This observation is consistent with experiment [28] and has also been observed computationally by Snook and van Megan [23] and Magda et al [29].

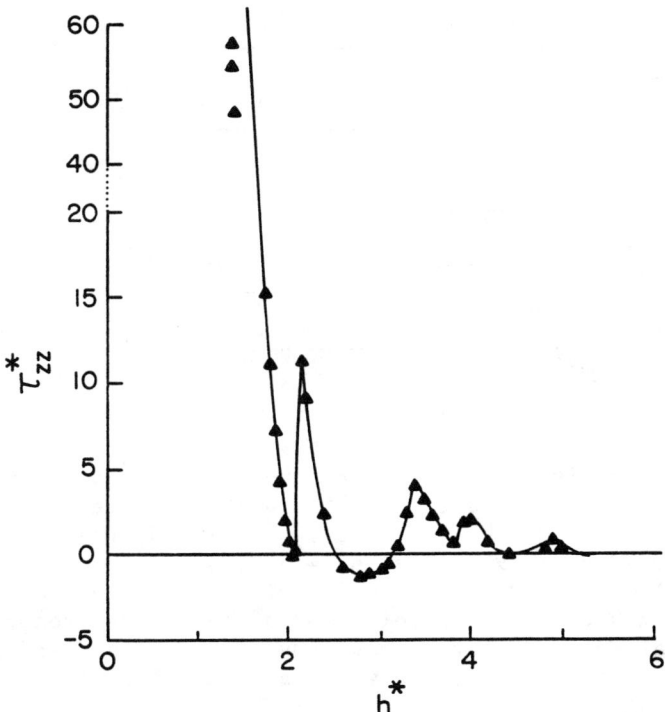

Figure 5. Plot of τ^*_{zz} vs h^* for data of Table 2 as computed by GCEMC.

Recently experiments have also been conducted on thin films undergoing shearing induced by transverse movement of the pore walls [30,31]. These measurements indicate that the solid surfaces slide past one another while separated by discrete numbers of fluid layers and that under appropriate conditions a critical shear stress, S_c, is required to initiate sliding. Moreover, S_c is again quantized with the number of fluid layers. The fact that a critical stress is needed to initiate sliding suggests that the pore fluid has assumed a solid-like structure that must be broken down in order for sliding to occur. We tested this hypothesis using our GCEMC and MD procedures [32].

Before discussing the shear-stress hypothesis we briefly review some data on the structure of the pore fluid. In Rhykerd et al. [33] we examined the structure of the pore fluid as a function of registry α. From Fig. 6 it is apparent that the pore fluid is quite sensitive to this parameter. In fact, for certain pore widths, it is possible alternately to freeze and liquify the "fluid" by changing α. This is most clearly illustrated by comparing the in-plane correlation function,

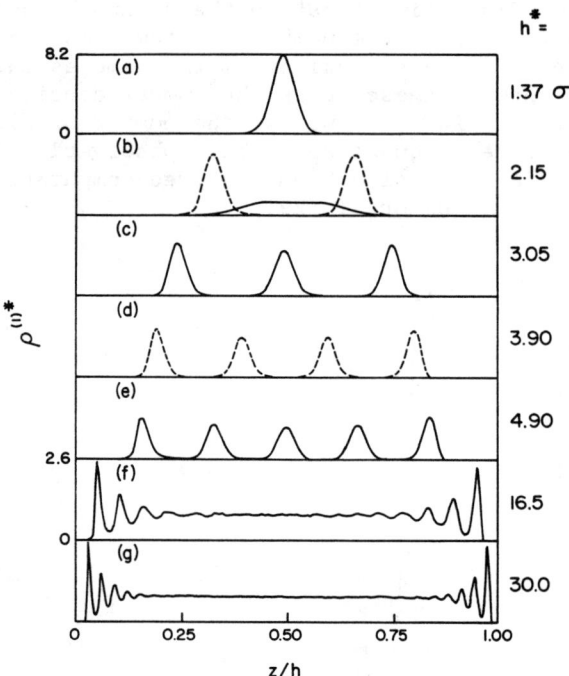

Figure 6. Local density profiles as a function of h^* at $T^* = 1.0$. The surface density of wall atoms is $N_s/s^{*2} = 0.7827$, where $N_s = 50$. Panel (a) $h^* = 2.37$, $\mu^* = -9.0$; (b) $h^* = 2.20$, $\mu^* = -9.26$; (c) $h^* = 3.05$, $\mu^* = -9.6$; (d) $h^* = 3.9$, $\mu^* = -9.83$; (e) $h^* = 4.9$, $\mu^* = -10.0$; (f) $h^* = 16.5$, $\mu^* = -10.29$; (g) $h^* = 30.0$, $\mu^* = -10.29$. In panels a-e the scale of $\rho^{(1)*}$ ranges from 0 to 8.2; the solid and dashed curves respectively correspond to walls in and out of registry. In panels f and g the scale of $\rho^{(1)*}$ ranges from 0 to 2.6; the walls are out of registry. From Rhykerd et al [30].

$g^{(2)}$, in Fig. 7 with the number density $\rho^{(1)}$ in Fig. 6. A sharply peaked $g^{(2)}$ indicates a solid-like structure whereas the liquid state is characterized by a much smoother function. As indicated by peaks in $\rho^{(1)}$, the fluid near the walls is layered parallel to the walls. For wider pores ($h^* = 16.5$, 30) the fluid is not layered near the pore center. However, the density of the fluid in this region is slightly less than that of a bulk phase (not shown) in equilibrium with the pore phase. This important point will be discussed later when we consider diffusion. At certain key values of α and h^* the pore fluid can be frozen even though it is in equilibrium with a bulk phase liquid. However, the solid structure may be destroyed by changing α. This is best illustrated in the second panel of Figs. 6 and 7. When $\alpha = 0.5$ the pore fluid is frozen; however, when α is changed to 0.0, a layer of fluid leaves the pore and the remaining pore fluid liquifies. This

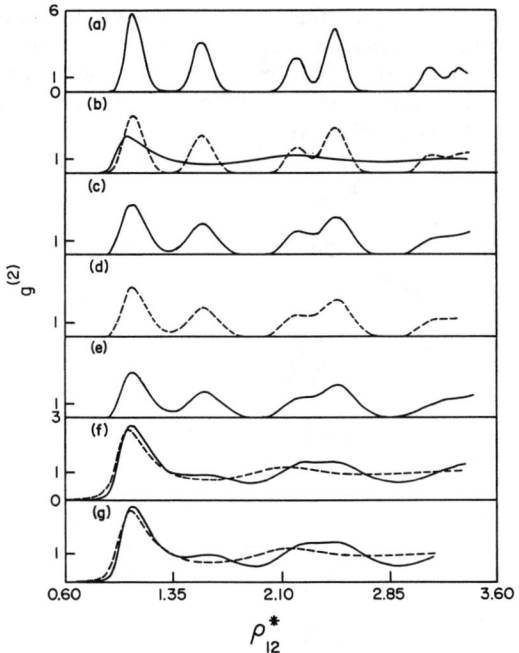

Figure 7. In-plane correlation functions corresponding to density profiles of Fig. 6. In panels a-e, only $g^{(2)}$ for the contact layer is shown, as $g^{(2)}$ values for the inner layers are indistinguishable from it; the scale of $g^{(2)}$ ranges from 0 to 6. In panels f and g, solid and dashed curves correspond respectively to contact and first inner layers; the scale of $g^{(2)}$ ranges from 0 to 3. From Rhykerd et al. [30].

implies the thermodynamic state space is not just a function of three parameters (e.g. μ,h,T), but rather four (e.g. μ,h,T,α). We surmise that a state of stable mechanical equilibrium obtains when there is no shear stress on the pore walls. If h^* is held fixed while wall 2 is slid relative to wall 1 in the x-direction (by changed α), then a transverse force should arise, tending to restore wall 2 to its equilibrium position.

To investigate this hypothesis, we have computed [32] τ_{zx} as a function of α for several values of h^*, all at the same fixed μ^* and T^*. Figure 8 displays plots of τ_{zx} versus α for three values of h^*. Because τ_{zx} is periodic in α, with a period of unity, only plots in the range -0.5 $\leq \alpha \leq$ 0.5 are shown. By virtue of the symmetry of the walls, τ_{zx} is antisymmetric in α. The shapes of the stress curves are similar, but the curve for h^* = 2.2 differs in phase by $\Delta\alpha$ = 0.5 from those for h^* = 3.1 and 4.9. The stress-strain curves have the following features. Both the slope of the linear region (the "force constant") and the critical stress decrease as the number of solid layers increases. However, the range of α over which solid-like behavior persists increases with the number of layers, that is, the critical strain (strain at which the solid changes to liquid) increases with the number of layers. The critical strain per layer is constant at 0.08. With increasing h*, the stress curve becomes flatter. At sufficiently large h*, the pore should become mechanically stable at all registries. The decrease of critical stress with increasing number of layers can be rationalized as follows. As the number of layers in the pore increases, the solid-like character of the layers decreases toward the center of the pore and it takes less force to break down the less ordered structure in the inner layers. All of the above observations are consistent with known experimental results.

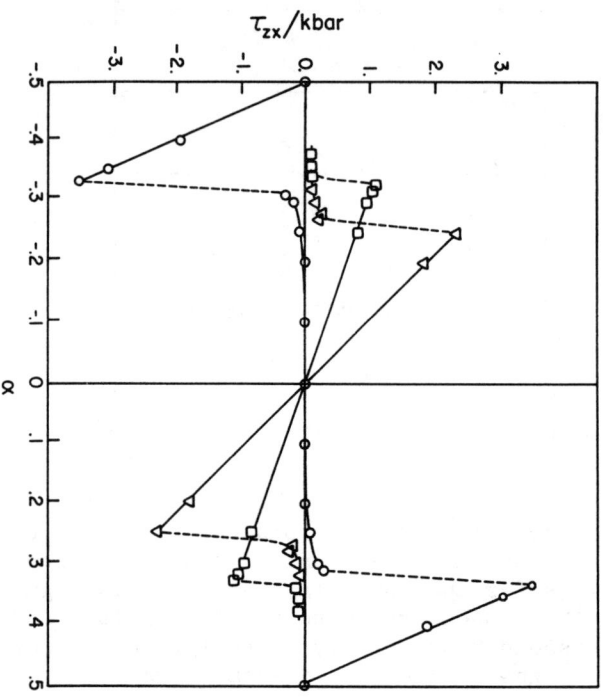

Figure 8. Shear stress τ_{zx} as a function of strain α for walls separated by h* = 2.2 (O), 3.10 (∇) and 4.9 ($\square$); μ^* = -9.26; T^* = 1.0. From Schoen et al. [29].

10. **Diffusion Results.** In this section we concentrate on the following aspects of diffuison: the validity of the continuum approximation and the classical expression for the diffusion tensor; diffusion parallel to the pore walls as a function of distance from the walls; dependence of diffusion on registry of the walls.

Typical experimental studies of pore fluids are performed with fixed μ, T and V(h). However, the MD simulations are carried out for fixed N, E, and V(h). Consequently each of our MD runs corresponds to a prior GCEMC run. The number of atoms used in the MD run is taken to be the average number <N> computed in the GCEMC run. The energy, E, of the system is adjusted until T in the MD run equals T in GCEMC. By invoking the "equivalence of ensembles" assumption of Sec. 2, μ in the MD run equals μ in the corresponding GCEMC run. By relying on an ergodic assumption, as metnioned in Sec. 2, we may compute expected values as time averages for known (μ, V, T).

Recall from Sec. 3 that the autocorrelation function, ψ, of the dynamical variable $\exp(i\mathbf{k}\cdot\mathbf{x})$ determines a unique representation for a generalized diffusion tensor that is valid at all wave-vectors and frequencies. This particular ψ is often called the self-part of the intermediate scattering function and is usually denoted by $F_s(\mathbf{k},t)$. To test the validity of the continuum approximation [7] of diffusion normal to the pore walls, we first compute $F_s(\mathbf{k},t)$ for small values of $|\mathbf{k}|$ characteristic of neutron-scattering experiments [34]. Given F_s, which is computed by MD for $\mathbf{k} = k\,\mathbf{e}_z = \pi h^{-1}\,\mathbf{e}_z$, we may determine $\tilde{D}_{zz}(k,\omega)$ from (33), (34) and (35). Figure 9 depicts $\tilde{D}_{zz}(k\mathbf{e}_z, \omega)$ and $\tilde{C}_z(\omega)$ for $h^* = 4.8$, 3.7 and 2.9 when $(\mu^*, T^*, \alpha) = (-10., 1., 0.5)$. By application of (49) we see that if the pore width is greater than approximately 3 atomic diameters, the continuum approximation holds. That is, for pore widths greater than 3σ the Fickian approximation is valid. Given the validity of Fick's law, we next test the applicability of (18) for diffusion normal to the pore walls. We know from (22) that for "large" pores (18) holds. The question is, how large is "large"? To answer this question we fit (20) in the least-squares sense to the MD-generated mean-square displacements $<(\Delta z)^2(t)>$. A typical fit is shown in Fig. 10. Again, we have found for pore widths greater than about 3σ the classical expression, (18), is valid.

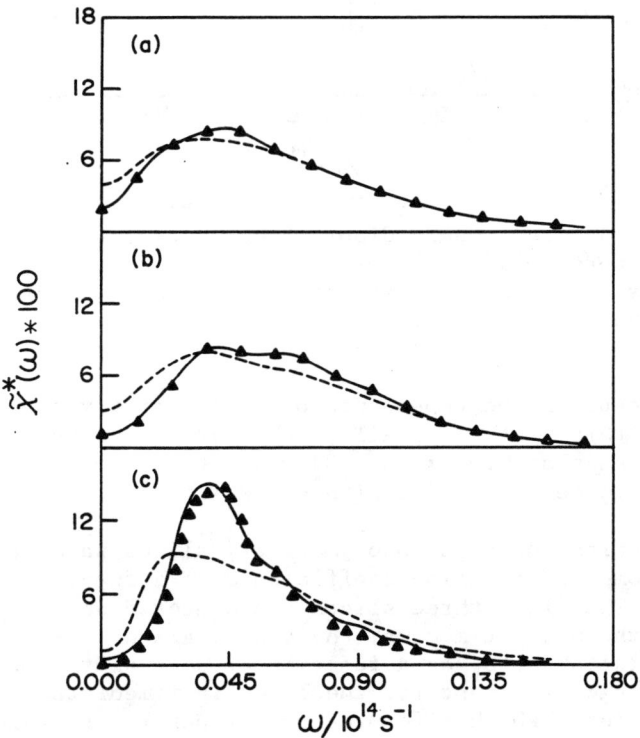

Figure 9. Fourier transforms of $D_{zz}(k\,\mathbf{e}_z,t)$ (——), $C_z(t)$ ($\triangle$), and $C_x(t)$ (- - -) for (a) $h^* = 4.8$, (b) $h^* = 3.7$, (c) $h^* = 2.9$. Here $\mu^* = -10.0$, $T^* = 1.0$, $\alpha = 0.5$ and $k = \pi/h$. From Schoen et al. [5].

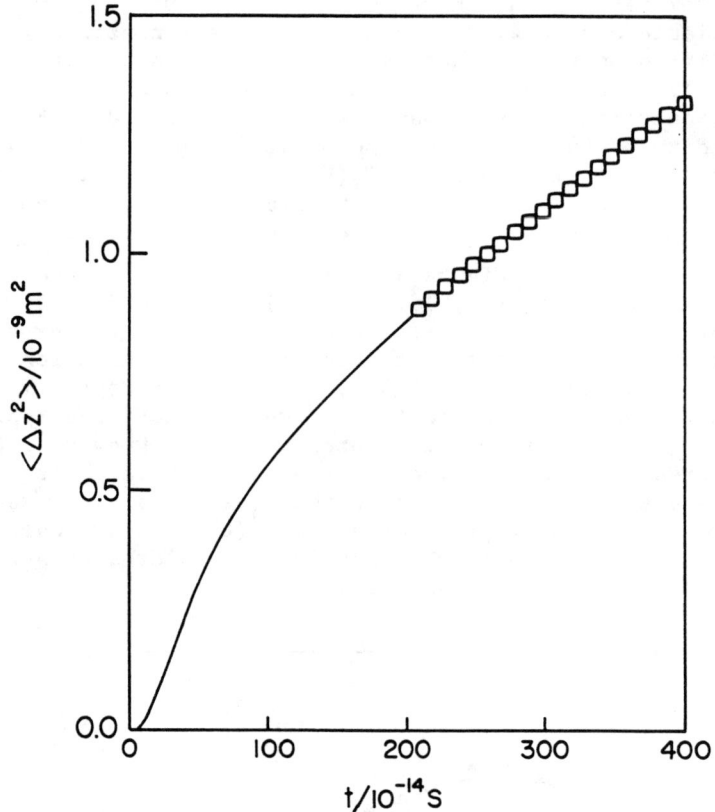

Figure 10. Plot of mean square displacement normal to the walls for the prototypal slit-pore at $\mu^* = -10.0$, $T^* = 1.0$, $h^* = 4.8$, and $\alpha = 0.5$. The open squares are from the least-squares fit of Eq. 20 using $j = 10$ terms. From Schoen et al. [5].

Table III depicts the dependence of D on registry for a pore of width 3.8σ. At this width, for all α, the pore contains 3 layers of fluid. The increase of $D(D_\perp = D_{zz}$, $D_\parallel = D_{xx} = D_{yy})$ with increasing α does not appear to be correlated with n^*, which remains essentially constant.

By partitioning the pore into imaginary slices parallel with the walls, we can compute diffusion coefficients as a function of distance from the wall. The first three slices are centered on the peaks of $\rho^{(1)}$ and bounded by the nearest minima; the others are arbitrary. Figure 11 depicts the variation of $D_\parallel$ as a function of pore width for fixed μ, α, T. The diffusion coefficient for the layer in immediate contact with the wall is smaller than that for the next inner layer, which is smaller than that for the next, and so forth. The most startling aspect of the figure is that the diffuison coefficient in the middle of the pore, for wide pores ($h^* = 16.5$, 30.0), is larger than in the bulk phase (which is in equilibrium with the pore phase). This enhanced diffusion, which has intriguing practical implications, was first noted in Schoen et al. [7].

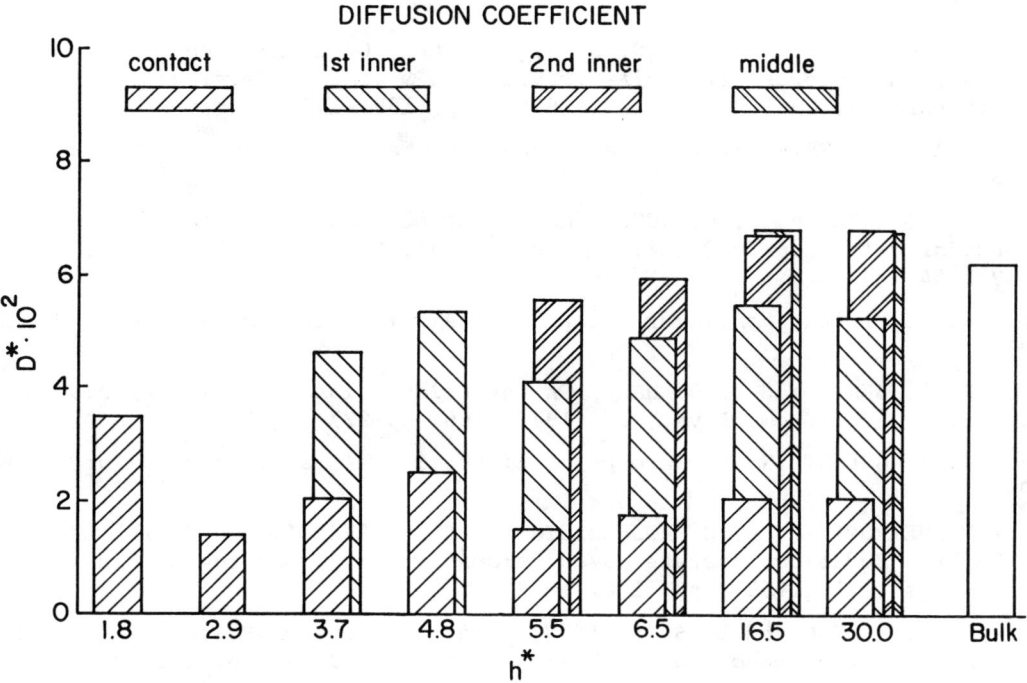

Figure 11. Self-diffusion coefficients in the slit-pore as a function of pore width at $\mu^* = -10.0$, $T^* = 1.0$, $\alpha = 0.5$. All values of D^* are multiplied by 10^2. Adapted from Table V of Schoen et al. [5].

Acknowledgements. The authors thank C. L. Rhykerd, Jr. for preparing Figures 5 and 11. They are grateful also to the Scientific Council of the Höchstleistungrechenzentrum at Forschungsentrum Jülich for a generous grant of computer time on the CRAY Y-MP/832 and to the Purdue University Computing Center for time on the CYBER 205 and ETA 10. M.S. is especially grateful to Professor Harald Morgner for his constant support. J.H.C. thanks R.K. Miller for valuable comments. Support for this research was provided by grants DE-FG02-86ER-60310 of the USDOE and DAAL03-90-G-0074 of the USARO.

References
1. J. P. BOON and S. YIP, *Molecular Hydrodynamics*, McGraw-Hill, New York, 1980.

2. M. P. ALLEN and D. J. TILDESLEY, *Computer Simulation of Liquids*, Oxford, Oxford, 1987.

3. P. F. LOW, *Structural component of the swelling pressure of clays*, Langmuir, 3 (1987), pp. 18-25.

4. R. ZWANZIG, *Incoherent inelastic neutron scattering and self-diffusion*, Phys. Rev., 133A (1964), pp. 50-51.

5. D. RUELLE, *Statistical Mechanics*, Benjamin, Reading, 1983.

6. M. REED and B. SIMON, *Functional Analysis*, Vol. 1, Academic, New York, 1972.

7. M. SCHOEN, J. H. CUSHMAN, D. J. DIESTLER, C. L. RHYKERD, Jr., *Fluids in micropores. II. Self-diffusion in a simple classical fluid in a slit-pore*, J. Chem. Phys., 88 (1988), pp. 1394-1406.

8. D.A. MCQUARRIE, *Statistical Mechanics*, Harper and Row, New York, 1976.

9. P. L. HALL and D. K. ROSS, *Incoherent neutron scattering for molecular diffusion in lanellar systems*, Mol. Phys., 36 (1978), pp. 1549-1554.

10. R. K. MILLER, *Nonlinear volterra integral equations*, Benjamin, NY, 1971.

11. B. J. BERNE and G. D. HARP, *On the calculations of time-correlation functions*, Adv. Chem. Phys., 17 (1970), pp. 63-227.

12. J. A. GOLDSTEIN, *Semigroups of Linear Operators*, Oxford, Oxford, 1985.

13. N. METROPOLIS, A. W. ROSENBLUTH, M. N. ROSENBLUTH, A. H. TELLER and E. TELLER, *Equation of state calculations by fast computing machines*, J. Chem. Phys., 21 (1953), pp. 1087-1092.

14. G. E. NORMAN and V. S. FILINOV, *Investigations of Phase Transitions by a Monte Carlo Method*, High. Temp. (USSR), 7 (1969), pp. 216-222.

15. D. J. ADAMS, *Grand canonical ensemble Monte Carlo for a Lennard-Jones fluid*, Mol. Phys., 29 (1975), pp. 307-311.

16. L. VERLET, *Computer 'experiments' on classical fluids. I. Thermodynamical properties of Lennard-Jones molecules.* Phys. Rev., 159 (1967), pp. 98-103.

17. R. VOGELSANG, M. SCHOEN and C. HOHEISEL, *Vectorization of molecular dynamics FORTRAN programs using the Cyber 205 vector processing computer*, Comp. Phys. Comm., 30 (1983), pp. 235-241.

18. M. SCHOEN, D. J. DIESTLER and J. H. CUSHMAN, *Fluids in micropores. I. Structure of a simple classical fluid in a slit-pore.* J. Chem. Phys., 87 (1987), pp. 5464-5476.

19. R. C. TOLMAN, *The Principles of Statistical Mechanics*, Dover, New York, 1979.

20. J. H. THURTELL and G. W. THURTELL, *Adsorption and diffusion at rough surfaces: a comparison of statistical mechanics, molecular dynamics, and kinetic theory*, J. Chem. Phys., 88 (1988), pp. 6641-6646.

21. J. P. VALLEAU, D. J. DIESTLER, J. H. CUSHMAN, M. SCHOEN, A. W. HERTZNER, M. E. RILEY, *Comment on "Adsorption and diffusion at rough surfaces: A comparison of statistical mechanics, molecular dynamics and kinetic theory*, J. Chem. Phys., (1990) (In Press).

22. W. F. SAAM and M. W. COLE, *Excitation and thermodynamics for liquid-helium films*, Phys. Rev. B, 11 (1975), pp. 1086-1105.

23. I. K. SNOOK and V. van MEGEN, *Physical adsorption of gases at high pressure. II. Effect of temperature.* Mol. Phys., 47 (1982), pp. 1417-1428.

24. N. QUIRKE, *Molecular simulation: progress and prospects*, Fluid Phase Equilib., 29 (1986), pp. 283-306.

25. B. K. PETERSON and K. E. GUBBINS, *Phase transitions in a cylindrical pore. Grand-canonical Monte Carlo, mean-field theory and the Kelvin equation*, Mol. Phys., 62 (1987), pp. 215-226.

26. M. SCHOEN, C. L. RHYKERD, Jr., J. H. CUSHMAN and D. J. DIESTLER, *Slit-pore sorption isotherms by the grand-canonical Monte Carlo method*, Mol. Phys., 66 (1989a), pp. 1171-1182.

27. K. L. CHUNG, *Markov Chains with Stationary State Probabilities*, Vol. 1, Springer, Heidelberg, 1960.

28. R. M. PASHLEY and J. N. ISRAELACHVILLI, *Molecular dynamics of water in thin films between mica surfaces and its relation to hydration forces*, J. Coll. Inter. Sci., 101 (1984), pp. 511-523.

29. J. J. MAGDA, M. TIRRELL, H. T. DAVIS, *Molecular dynamics of narrow, liquid-filled pores*, J. Chem. Phys., 83 (1985), p. 1888.

30. J. N. ISRAELACHVILLI, P. M. MCGUIGGAN and A. M. HOMOLA, *Dynamic properties of molecularly thin liquid films*, Science, 240 (1988), pp. 189-191.

31. J. van ALSTEN and S. GRANICK, *Molecular tribology of ultra thin liquid films*, Phys. Rev. Lett., 61 (1988), pp. 2570-.

32. M. SCHOEN, C. L. RHYKERD, Jr., D. J. DIESTLER and J. H. CUSHMAN, *Shear forces in molecularly thin films*, Science, 245 (1989b), pp. 1223-1225.

33. C. L. RHYKERD, Jr., M. SCHOEN, D. J. DIESTLER and J. H. CUSHMAN, *Epitaxy in simple classical fluids in micropores and near surfaces*, Nature, 330 (1987), pp. 461-463.

34. M. NIELSEN, J. P. MCTAGUE and L. RUSSELL, *Neutron-scattering studies of physiosorbed monolayers on graphite*, in Phase Transitions in Surface Films, J. G. Dash and J. Ruvalds, eds., NATO-ASI, 51, Plenum, New York, 1979, pp. 127-163.

Table I.   Reduced variables

$$r^* = r/\sigma \qquad\qquad n^* = \langle N \rangle \, \sigma^3/(s^2 h)$$

$$h^* = h/\sigma \qquad\qquad T^* = k\, T/\varepsilon$$

$$A^* = A/\sigma^2 \qquad\qquad \mu^* = \mu/\varepsilon$$

$$V^* = V/\sigma^3 \qquad\qquad \rho^* = \rho^{(1)} \, \sigma^3$$

$$\tau^*_{xz} = \tau_{xz} \, \sigma^3/\varepsilon \qquad\qquad D^* = D/(\varepsilon\sigma^2/m)^{1/2}$$

$$\tau^*_{zz} = \tau_{zz} \, \sigma^3/\varepsilon \qquad\qquad s^* = s/\sigma$$

$$U^* = U/\varepsilon$$

Table II   Data for Figure 5

h^*	τ_{zz}^*	$-\mu^*$	α
1.37	57.14	9.00	0.0
1.38	54.02	9.00	0.0
1.40	47.77	9.00	0.0
1.75	15.26	9.03	0.5
1.80	11.12	9.06	0.5
1.85	7.20	9.08	0.5
1.90	4.27	9.11	0.5
1.95	2.06	9.14	0.5
2.00	0.70	9.16	0.5
2.05	-0.05	9.19	0.5
2.10	0.16	9.21	0.5
2.15	11.37	9.23	0.5
2.20	9.15	9.26	0.5
2.40	2.34	9.34	0.5
2.60	-0.68	9.42	0.5
2.80	-1.21	9.50	0.5
2.90	-1.12	9.53	0.5
3.00	-0.77	9.57	0.5
3.10	-0.36	9.60	0.5
3.20	0.56	9.63	0.5
3.30	2.51	9.66	0.5
3.40	4.03	9.69	0.5
3.50	3.27	9.72	0.5
3.60	2.21	9.75	0.5
3.70	1.26	9.78	0.5
3.80	0.80	9.80	0.5
3.90	1.94	9.83	0.5
4.00	2.10	9.85	0.5
4.20	0.80	9.90	0.5
4.40	0.15	9.95	0.5
4.80	0.59	10.03	0.0
4.90	1.05	10.00	0.0
5.00	0.62	10.07	0.0

Table III

Self-diffusion coefficient as a function of registry α for the slit-pore at $\mu^* = -10.0$, $T^* = 1.0$, and $h^* = 3.8$. Values of D^* are multiplied by 10^2.

α	n^*	D^*_{xx}	D^*_{zz}
0.5	0.6179	3.39	1.40
0.33	0.6118	3.48	1.37
0.25	0.6082	3.11	1.13
0.17	0.6112	2.58	0.92
0.0	0.6027	2.55	0.76

CHAPTER 6

Non-Linear Dynamics in Chemically Compacting
Porous Media**

T. Dewers*
P. Ortoleva*†

Abstract. The consolidation or densification of porous media in the
later stages of burial can occur by stress-induced mass transfer from
grain contacts by a process termed "chemical compaction" or "pressure
solution". We examine the consequences of chemical compaction in
texturally heterogeneous rock by placing a grain-to-grain interaction
model within a continuum formalism capturing the dynamics of reaction,
diffusion, and the mechanics of fluid-pressured porous media. The nonlinear and coupled nature of the equations which constitute the model
presents interesting possibilities for phenomena common to other nonlinear dynamical systems. Numerical simulations are presented; examples
show the spatial self-organization of banded compaction/cementation
couplets in sandstones, the development of spatially localized
compaction/dissolution zones termed "stylolites", and a temporal
oscillatory cycle of dissolution and plastic deformation within
stylolitic zones.

Introduction. The compaction of rocks in a sedimentary column
during burial proceeds by both mechanical and chemical processes.
Depending on burial rate, solely mechanical processes dominate in the
upper kilometer or so of the earth's crust, while chemical compaction,
usually termed "pressure solution", becomes dominant in the later

stages of burial.[1,2] In what follows we show that the coupling of rock
mechanics and chemistry leads to a number of pattern formation effects
that accompany compaction in sedimentary rocks. These are not only

---

*Departments of Geological Sciences and, +Chemistry, Indiana University,
Bloomington, IN 47405.
** Research supported in part by a contract from the Gas Research
Institute, a grant from the Earth Sciences Division of the National
Science Foundation and the Basic Energy Sciences Program of the US
Department of Energy. Some numerical simulations were carried out on an
IBM RT.

interesting in themselves in the context of nonlinear dynamics but also are important from the perspective of oil and gas exploration and reservoir engineering. We first review some of these phenomena. In later sections we present quantitative models and numerical simulations that illustrate temporal oscillation and spatial self-organization in the spatio-temporal evolution of rocks under applied stress and undergoing grain growth/dissolution reactions.

Stylolites. Stylolites are roughly planar surfaces or seams in nominally monomineralic rocks into which adjacent rock has converged and interpenetrated.[3,4] It is thought that stylolite formation, or "stylolitization", involves some process of pressure solution which has become localized in space.[5] The plane of stylolitization is generally believed to be oriented normal to the direction of maximum principle compressive stress, so that stylolitization is somehow driven by stress.[6] As it involves the dissolution of grains within the core of the stylolite and commonly redeposition at the sides, stylolitization likely includes some reaction-diffusion process.[7]

In the modeling to be presented here, we show how stylolites result from an instability of a chemically compacting rock to localized compaction by mechano-chemical feedback. Because of the ellipsoidal shape of detrital grains and the common development of overgrowth at adjacent free faces, dissolution at grain contacts will result in an increased contact area. For a constant applied far-field stress, this decreases the stress, or force per area, across the contacts. This is a stabilizing effect, as any region with a slightly smaller contact area than that in its surroundings will compact faster, thus driving the contact area to be in line with that in its surroundings. If, however, the contact area can be prevented from increasing in a spatial locality, compaction may proceed in that locality at a higher than average rate. Two mechanisms which can prevent the increase in contact area accompanying chemical compaction include simultaneous free-face dissolution (i.e. dissolution at grain faces in contact with pore fluid) and the prevention of overgrowth formation by the presence of clay coatings.[8]

The synchronous operation of both contact and free-face dissolution has been advocated in the geologic literature.[9,10] We conjecture that the strain energy characteristic of solid-fluid interfaces is affected by the average state of stress in the grain, while that at grain contacts is more influenced by the intergranular stress acting across grain contacts.[8] The average stress within a grain is an increasing function of porosity in that locality.[7,11,12] Thus free-face dissolution, as it increases the "porosity" surrounding a grain, auto-focuses the average strain energy in space by positive feedback involving dissolution at regions slightly higher in porosity, transport to adjacent regions of lower porosity (and thus lower in average free energy), and reprecipitation there.[7,8] Between two adjacent regions of slightly different porosity, as the contrasts in texture increase with differential dissolution and cementation, so does the free energy gradient driving it.

The occurrence of either contact-dissolution mediated (stable) compaction versus (unstable) free-face dissolution depends on the relative rates of both processes. We have observed a transition from stable to unstable compaction by, among other things, increasing the ratio of rate coefficients for free face and contact rate laws.[8]

Banded Compaction-Cementation Alternations. In addition to stylolitic compaction, repetitive, banded alternations of heightened compaction and cementation, in which the compaction zone is wider than in the case of stylolitization, are observed to occur in sandstones[13,14] and in marl/limestone sequences.[15] Stylolites may or may not be associated with these mm-to-cm wide zones; when they are, they occur at boundaries between compaction and cementation zones.[15] The spatial segregation between dissolution and cementation zones is thought to occur by mechano-chemical feedback similar in nature to that for stylolites described above.[8]

For a constant applied (far-field) stress, the intergranular stress (i.e. that across grain contacts) and thus the chemical potential at contact facets, increases with decreasing contact area. Two regions of contrasting contact area thus may have a chemical potential gradient existing between them. Small variations in contact radii can auto-enhance because mass lost at grain contacts in regions of relatively smaller contact area can precipitate on other grains in regions of relatively greater contact area and not necessarily (as is commonly favored in pressure solution models) on the free faces on the same grain adjacent to contacts. If the contact area increase due to overgrowth cementation relative to the contact area increase accompanying compaction is such that the contact area difference is maintained between the two spatial regions, the chemical potential gradient existing between the regions is maintained.

In the later stages of chemical compaction, overgrowth formation results in a greater degree of porosity occlusion, and thus contact area increase, than does compaction (for a given amount of mass lost at contacts and gained at free faces).[16] In this case, the chemical potential gradient and related reaction and transport between the two regions is self-sustaining. We separate conceptually this behavior from that of stylolitization as the latter may involve dissolution at free faces.[8] As will become clear later, the free-face dissolution domain of behavior is more "unstable", leading to a more highly localized zone of compaction.

Temporally Oscillatory Compaction. While spatial patterns of mineralization may be relatively easily identified either directly or via relic patterns, oscillatory temporal behavior on geologic time scales can generally only be inferred. Here we present plausibility arguments for temporally oscillatory effects that arise as a consequence of nonlinear mechano-chemistry.

The "stylolitization" instability as viewed in this paper involves free face dissolution at grain contact peripheries which, by itself, acts to decrease contact areas. If increasing intergranular stress resulting from the decreasing contact area exceeds the compressive strength of the mineral constituting the contact, plastic deformation or breakage due to micro- fractures at the grain-grain contact will occur, bringing the contact area to a level such that the intergranular stress equals the compressive strength.[17] The aforementioned stylolite generation mechanism might then continue until plastic deformation again sets in. Thus we expect that if the rates of plastic deformation and dissolution are roughly similar, compaction and dissolution may occur by an oscillatory dynamic.

These observations and qualitative arguments suggest that the coupling of mechanical, reaction and transport effects in rocks under stress can yield a richness of nonlinear dynamical behavior. In the present work, we focus on the instability of the uniform state of compaction to the formation of banded textural patterns and temporal oscillation at runaway sites of very localized compaction. Our objective is to develop quantitative mechanical-reaction-transport models capable of predicting the properties and range of existence of these phenomena.

Several factors make the study of nonlinear phenomena in compacting, reacting media challenging. Firstly, these problems are by nature transient in that once the medium reaches zero porosity, further compaction is not possible. Furthermore, as grains do not "diffuse" (i.e. migrate relative to each other) the linear stability analysis shows that unstable modes driving pattern formation either show no particular preferred wavelength or even if they do, very short wavelength perturbations of all wavelengths down to the grain size (where a continuum theory breaks down) are amplified.[18] These two cases (referred to as svegliable and cognito respectively[19]) are common features in patterning in precipitate systems (or more generally any reaction-transport problem where some of the dynamical variables do not have a diffusional dynamics). Because of these difficulties we have used numerical simulations of our mechanical-reaction-transport model in the present work to demonstrate the nonlinear phenomena in these systems. Neither a linear stability analysis nor some generalization of bifurcation theory for transient systems were attempted.

In summary the nonlinear dynamics of mechano-chemical systems presents many challenging mathematical problems. These include transient bifurcations and the theory of coupled systems involving partial differential equations of various types and algebraic equations. Advances made in analysis and numerical computation of these problems will be of interest in both academic and industrial communities.

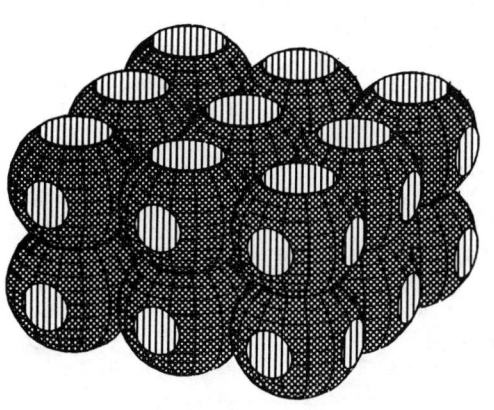 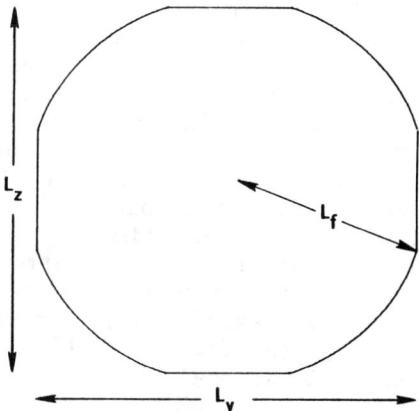

Figure 1. Textural model used to quantify grain-scale processes. The "rock" consists of a periodic array of truncated spheres in rectangular packing. L_f is the radius of the spheres; L_z, L_x and L_y form the "box" truncating the sphere. Porosity, grain number density, and the various surface areas for grain-grain contacts and free faces are determinable from L_x, L_y, L_z and L_f.

MECHANICAL-REACTION-TRANSPORT MODEL. Modeling chemical compaction involves combining elements of rock mechanics, the thermodynamics of stressed solids, crystal growth, and the kinetics of reaction and transport in an aqueous fluid-saturated porous medium.[8] As such, we expect the mathematical structure of the theory to be rather complicated, involving the simultaneous solution of equations of mass balance, force balance and crystal growth. In this section we present a model for the simplest case of a single load-bearing mineral and further restrict our considerations to one spatial dimension. A number of aspects of the theory required to relax these restrictions are under investigation in our laboratory.

Textural Model and Kinematic Relations. Rocks are composite media made up of individual grains that adhere to form an aggregate solid. The "texture" Θ is a term applied to that collection of variables describing the grains. In the simplest case Θ could be average grain volume and number of grains per rock volume for each mineral and the porosity ϕ (volume fraction of the rock occupied by pore space). In other cases more complex textural models are required. Additional variables often encountered are grain shape and orientation and the probability distribution for these quantities.

Compaction (porosity loss) commonly accompanies a, predominant shortening in one direction – typically vertical. In these cases grain height is a key variable. If we think of the rock as being constituted of an array of truncated spherical grains, Weyl's[12,20] grain geometric model shown in Fig. 1, then

* compaction in the z-direction corresponds to a shortening of the L_z variable;
* grain overgrowth corresponds to an increase of the radius L_f of the truncated sphere; and
* the grains come in contact with the pore fluid at the spherical sections (denoted "free faces").

Clearly, real rocks are not a periodic array of such grains but we use the above textural model to suggest the vertical grain dimension L_z, the grain radius L_f and the two lateral grain dimensions L_x and L_y (see Fig. 1). More complex packing models have also been used. The advantage of the Weyl model is that stresses at grain contacts are easily calculable. For the present study we assume the horizontal dimensions L_x and L_y to be constant.

Consider now evolution of the above texture variables in a one dimensional system along the (vertical) z axis. Then L_z and L_f are functions of z and time t. Let G_z be the rate of change of L_z due to reaction at the top and bottom grain-grain contacts and due to plastic deformation and u be the velocity of rock deformation (compaction) along the z axis. Then in a small time δt we have

$$L_z(z+u\delta t, t+\delta t) = L_z(z,t) + G_z \delta t . \tag{1}$$

Because δt is small but arbitrary this implies

$$\frac{\partial L_z}{\partial t} = -u\frac{\partial L_z}{\partial z} + G_z . \tag{2}$$

Similarly,

$$\frac{\partial L_f}{\partial t} = -u\frac{\partial L_f}{\partial z} + G_f . \tag{3}$$

If we neglect nucleation, conservation of grain number density n implies

$$\frac{\partial n}{\partial t} = -\frac{\partial(un)}{\partial z} . \tag{4}$$

But noting that $L_x L_y L_z n=1$ we may show that

$$\frac{\partial u}{\partial z} = \frac{G_z}{L_z} \tag{5}$$

by using (2,4). It follows that the coupling between the dynamics of the textural variables L_z and L_f, the pore fluid composition and the stresses follow from the dependence of G_z and G_f on these factors.

Conservation of Solute Mass. Consideration of conservation of mass of the solute molecules in the fluid filling the pores yields reaction-transport equations for these concentrations. Here we present these equations and analyze them in the limits where: (1) reactions in the pore fluid are fast relative to grain growth/dissolution reactions, and (2) concentrations of solutes in the pores are much less than the molar density of solids constituting the grains.[21] Both these assumptions are well satisfied in most geological contexts of interest here. The usual reaction-diffusion equations must be supplemented by a number of terms accounting for flow through the medium and for grain growth/dissolution reactions at various distinct sites on the grains. We find[8]

$$\frac{\partial \phi c_\alpha}{\partial t} = -\frac{\partial J_\alpha}{\partial z} + \phi \sum_{k=1}^{N} \nu_{\alpha k} W_k + n \rho \mu_\alpha \sum_{i=z,f} A_i \hat{G}_i \ . \tag{6}$$

The net rate of change of moles of species α in the N solute system ($\alpha=1,2,...N$) per rock volume (the LHS) has three contributions. The first is due to transport, J_α being the flux (moles/area-time). The W_k are the rates (moles/volume-time) of each of the N_a aqueous reactions ($k=1,2,...N_a$); the $\nu_{\alpha k}$ are stoichiometric coefficients for these reactions; μ_α is a stoichiometric coefficient for α in the mineral dissolution reaction. The last term is the contribution from the two sites of mineral reaction – the free face (f) in contact with the pore fluid and the top and bottom grain-grain contacts (z); A_f is the area of the spherical part of the grain while A_z is the area of a z-contact. The A_i may be calculated in terms of L_x, L_y, L_z and L_f for the simple truncated sphere model. Finally, in (6) we distinguish G_i, the net rate of change of L_f due to both reaction and mechanical deformation, and $\hat{G}_i$, the part of G_i from reaction only.

In many geological applications the aqueous reactions are much faster than other processes (such as grain growth/dissolution). We express this by writing

$$W_k = E_k/\varepsilon \tag{7}$$

where E_k is a factor that vanishes when the k-th reaction is at equilibrium. In the limit of fast aqueous reactions, $\varepsilon \to 0$, Eqn. (6) is not in an appropriate form for numerical computation – i.e. small errors in departures of E_k from zero will be greatly amplified by the ε^{-1} factor. To surmount this we take linear combinations of equations of the form (6) for various values of α such that the W_k terms are eliminated. This may be accomplished by introducing row vectors $a^{(\lambda)} = \{a_1^{(\lambda)}, a_2^{(\lambda)}, ... a_N^{(\lambda)}\}$ which are orthogonal to all linearly independent columns of the matrix $\underline{\underline{\nu}}$:

$$\sum_{\alpha=1}^{N} a_\alpha^{(\lambda)} \nu_{\alpha k} = 0 \ . \tag{8}$$

The linearly independent columns of ν correspond to independent equilibria; for example, the reactions $X+Y=Z$ and $2X+2Y=2Z$ may be different kinetically but yield the same equilibrium condition. Introducing the notation

$$J^{(\lambda)} = \sum_{\alpha=1}^{N} a_{\alpha}^{(\lambda)} J_{\alpha} \tag{9}$$

$$\omega^{(\lambda)} = \sum_{\alpha=1}^{N} a_{\alpha}^{(\lambda)} \mu_{\alpha} \tag{10}$$

we then obtain

$$\sum_{\alpha=1}^{N} a_{\alpha}^{(\lambda)} \frac{\partial \phi c_{\alpha}}{\partial t} = -\frac{\partial J^{(\lambda)}}{\partial z} + n\rho\omega^{(\lambda)} \sum_{i=z,f} A_i \hat{G}_i . \tag{11}$$

These equations must be solved subject to the equilibrium conditions

$$E_k(\underline{c},p) = 0 \text{ for fluid pressure p.} \tag{12}$$

If there are N_a^* independent such relations then there are $N-N_a^*$ equations of the type (11), constituting a sufficient number (N) of equations to determine the N concentrations $c_1, c_2, \ldots c_N$.

Another factor that places strong constraints on the dynamics of the reactive porous media of interest is the relative insolubility of most minerals. A typical concentration (denoted c) is much less than the solid molar density ρ. This introduces a smallness parameter δ ($\equiv \bar{c}/\rho$) into the problem. We expect that the time to make appreciable changes in mineralization will be of order δ^{-1}. Furthermore, the $\hat{G}_i$ have embedded within them a factor ρ^{-1} and at least one factor of concentration (and hence $\bar{c}$). Also, the compaction velocity evolves on the time scale of grain dimension change. Hence we expect u, G_i, and $\hat{G}_i$ to be of order δ. With these arguments we make the following change of variables:

$$c_{\alpha} = \bar{c} c_{\alpha}', \quad G_i = \delta G_i', \quad \hat{G}_i = \delta \hat{G}_i' ,$$

$$\delta = \bar{c}/\rho \; (\ll 1), \quad t = \frac{1}{\delta} t' \tag{13}$$

$$J^{(\lambda)} = \bar{c} J^{(\lambda)\prime}, \quad u = \delta u' .$$

Placing these definitions into (11) and letting $\delta \to 0$ we obtain, neglecting the primes henceforth,

$$\frac{\partial J^{(\lambda)}}{\partial z} = \sum_{i=z,f} \omega^{(\lambda)} A_i \hat{G}_i \tag{14}$$

where (2,3,5) retain their forms.

Reaction Driving Forces and Quasi-Linearity. The formation of stylolites and banded compaction features involves dissolution of grains, mass transfer of solutes, and precipitation of these solutes at another point in space. This redistribution requires a driving force – the variation in the free energy of grain facets that follows from variations in texture in the stressed rock. That the variations in free energy are small relative to that implied by rock solubility itself suggests that variations of solute composition attending the differentiation processes will be small. This then ensures that under most conditions of interest here, the reaction-transport equations will be quasi-linear with respect to the deviation of concentrations from their average value. This is not only important in understanding the essential structure of the self-organization dynamics and identifying the central nonlinearity but also is of great computational advantage.

The part of the rates of change of the textural variables due to chemical reaction, $\hat{G}_i$, can for mass action kinetics be written in the form

$$\hat{G}_i = q_i[Q_i(\underline{c},p) - K_i] . \tag{15}$$

The factor q_i is positive definite and the condition for equilibrium with respect to site $i(=z,f)$ is written $Q_i-K_i = 0$. As mentioned above, we expect that the concentrations lie near equilibrium with the grains in some reference state and therefore the concentrations deviate from their typical values (defined by this equilibrium state) by only a small amount. More precisely, we introduce the following variables:

$$c_\alpha = \bar{c}_\alpha(p)[1+\gamma_\alpha] \qquad (16)$$

$$K_i = \bar{K}(p)[1+\kappa_i] \qquad (17)$$

where the reference "solubility" $\bar{K}(p)$ is independent of i (site type) but depends on fluid pressure p. The scaled deviations γ_α and κ_i contain the relevant variations for inducing mechano-chemical pattern formation. As they are small we neglect quadratic and higher order terms in these quantities. With this we obtain

$$\hat{G}_i \simeq \bar{q}_i\bar{K}\left[\sum_{\alpha=1}^{N}\Lambda_{i\alpha}\gamma_\alpha-\kappa_i\right] \qquad (18)$$

where $\bar{q}_i$ is q_i evaluated at $\bar{c}_1,\bar{c}_2,...\bar{c}_N$. The factors $\Lambda_{i\alpha}$ are given by

$$\Lambda_{i\alpha} = \overline{\partial Q_i/\partial c_\alpha} \qquad (19)$$

where the bar again indicates evaluation at $\bar{c}_1,\bar{c}_2,...\bar{c}_N$.

The equilibrium conditions (12) are also needed to complete the quasi-linear problem. By definition $\bar{c}$ satisfies

$$E_k(\bar{c},p) = 0 . \qquad (20)$$

Then placing (16) into (12) and only retaining terms linear in the γ's we obtain

$$\sum_{\alpha=1}^{N}e_{k\alpha}\gamma_\alpha = 0 \qquad (21)$$

where

$$e_{k\alpha} = \overline{\frac{\partial E_k}{\partial c_\alpha}}\,\bar{c}_\alpha . \qquad (22)$$

If transport is only by (diagonal) diffusion then

$$J_\alpha = -D_\alpha\frac{\partial c_\alpha}{\partial z} . \qquad (23)$$

With this (14) yields

$$\sum_{\alpha=1}^{N}a_\alpha^{(\lambda)}\,\frac{\partial}{\partial z}\left(\phi D_\alpha\frac{\partial \bar{c}_\alpha\gamma_\alpha}{\partial z}\right) + n\rho\omega^{(\lambda)}\sum_{i=z,f}A_i\bar{q}_i\bar{K}\left[\sum_{\beta=1}^{N}\Lambda_{i\beta}\gamma_\beta-\kappa_i\right] = 0. \qquad (24)$$

Thus the N_a^* linearized independent equilibrium conditions and the $N-N_a^*$ modified conservation of mass equations are sufficient to determine the N γ's. Finally, the $\hat{G}_i$ contributions in G_i on the RHS of (2,3) can be taken in their quasi-linear form (18).

Note that the above prescription does not completely fix the $\bar{c}_\alpha$ for most cases of geological interest. There are N_a^* equilibrium relations for the $\bar{c}_\alpha$. Also, because the two sites $i=z,f$ are for the same mineral, the term Q_i is independent of i (assuming the chemical mechanism to be the same at both sites). Hence the conditions $Q_i(\bar{c},p)$ = $\bar{K}(p)$ only yields one condition. Thus we have only given N_a^*+1 conditions to fix the N $\bar{c}_\alpha$. Hence we require $N-N_a^*-1$ additional

conditions. These are supplied from the "geochemistry" of the problem - we must give $N-N_a^*-1$ conditions on the $\bar{c}_\alpha$ that reflects, for example, whether the fluid is saline or fresh or whether the fluid is held in equilibrium with other solids or fluids (such as oil or natural gas). Specifically we can write the extra constraints as follows. One can choose the $a_\alpha^{(\lambda)}$ with the following extra conditions

$$\omega^{(\lambda)} = 0, \quad \lambda=1,2,\ldots N-N_a^* -2 \tag{25}$$

$$\omega^{(N-N_a^*-1)} \neq 0 \ . \tag{26}$$

Then the extra constraints needed can be written

$$\sum_{\alpha=1}^{N} a_\alpha^{(\lambda)} \bar{c}_\alpha = \psi^{(\lambda)}, \quad \lambda=1,2,\ldots N-N_a^*-2 \ . \tag{27}$$

Here the $\psi^{(\lambda)}$ must be specified according to the geochemical conditions of interest.

Equations For Macroscopic Effective Stress and Force Balance. The goal of the mechanical facet of the modeling is to fit the grain geometry (microscopic) picture into the continuum dynamics of textural variation on the supra-grain scale. We must calculate some approximation for the microscopic (grain scale) distribution of stresses because variations in stress between the grain-grain contact and that applied at the free faces in contact with the pore fluid underlies many of the phenomena of interest. In this section we present the partial differential equations necessary to describe elastic interactions as a function of applied stresses, fluid pressure, and attendant changes in the textural properties in the single stress-supporting mineral system. These take the form of three second-order Navier equations for the three components of macroscopic elastic displacement, w_x, w_y, and w_z. To derive these, we require a constitutive relationship between the macroscopic stress and strain tensors, a condition for force balance, and an expression relating the displacement vector with components of the strain tensor.

The macroscopic stress ($\underline{\underline{\sigma}}^m$) and strain ($\underline{\underline{\varepsilon}}^m$) tensors are taken to be related by a fourth rank tensor of elastic compliances $\underline{\underline{C}}^m$, fluid pressure p, and an effective stress coefficient α^m:[22]

$$\underline{\underline{\sigma}}^m + \alpha^m p \underline{\underline{I}} = \underline{\underline{C}}^m \underline{\underline{\varepsilon}}^m \qquad (\underline{\underline{I}} \text{ is the identity tensor}) \ . \tag{28}$$

In the case of an isotropic medium, the compliance tensor is assumed to be related to a macroscopic shear modulus μ^m and Lame's constant λ^m by

$$C_{ijkl}^m = \mu^m(\delta_{ik}\delta_{jl} + \delta_{il}\delta_{jk}) + \lambda^m\delta_{ij}\delta_{kl} \tag{29}$$

and α^m is related to the bulk modulus of both the porous solid skeleton, κ^m (drained of pore fluid) and the solid composing the skeleton, κ by[11,23]

$$\alpha^m = 1 - \kappa^m/\kappa. \tag{30}$$

κ^m is related to μ^m and λ^m (and similarly κ is related to μ and λ). A number of authors have derived upper and lower bounds for the isotropic moduli of porous media as a function of porosity.[24]

Force balance in the absence of body force implies

$$\sum_{j=1}^{3} \frac{\partial \sigma_{ij}^m}{\partial x_j} = 0 \ . \tag{31}$$

For linear elasticity, the strain tensor $\underline{\underline{\varepsilon}}^m$ and the vector of elastic displacements $\underline{w}$ are related by

$$\varepsilon_{ij} = \frac{1}{2}\left(\frac{\partial w_i}{\partial x_j} + \frac{\partial w_j}{\partial x_i}\right) \ . \tag{32}$$

Combining (28), (31) and (32) yields

$$\frac{\partial}{\partial x}\left\{(\lambda^m + 2\mu^m)\frac{\partial w_x}{\partial x} + \lambda^m\frac{\partial w_y}{\partial y} + \lambda^m\frac{\partial w_z}{\partial z} - \alpha^m p\right\} = 0$$

$$\frac{\partial}{\partial y}\left\{(\lambda^m + 2\mu^m)\frac{\partial w_y}{\partial y} + \lambda^m\frac{\partial w_x}{\partial x} + \lambda^m\frac{\partial w_z}{\partial z} - \alpha^m p\right\} = 0 \qquad (33)$$

$$\frac{\partial}{\partial z}\left\{(\lambda^m + 2\mu^m)\frac{\partial w_z}{\partial z} + \lambda^m\frac{\partial w_x}{\partial x} + \lambda^m\frac{\partial w_y}{\partial y} - \alpha^m p\right\} = 0 .$$

<u>Grain Reaction Rates: Mechanisms and Driving Forces.</u> Quantification of the rates of growth and dissolution at the various grain facets requires an account of both driving forces and rate limiting steps involved in the transfer of mass between solid and fluid phases at each facet type. As grain reaction reflects stresses at the grain scale, we need to account for the connection between texture, microscopic stresses, and the macroscopic stress field within the constraints imposed by our assumption of a single load-supporting phase and simple grain geometry.

The average vertical normal stress across grain contacts is found by equating the average force within a horizontal plane to the area-weighted average of forces in that plane acting within both solid and fluid. Let P_z be the normal pressure across the z contact (equal to minus the normal stress as per the sign convention of negative compressive stresses), and note that L_xL_y is the horizontal area of the "unit cell". It follows that[24]

$$-L_xL_y\sigma^m_{zz} = A_zP_z + (L_xL_y - A_z)p . \qquad (34)$$

Relations similar to (34) hold for P_x and P_y.

We assume that the average stress within grains is determined from the macroscopic mean stress (= $-\mathrm{tr}\underline{\underline{\sigma}}^m/3$) by averaging the contributions by fluid and solid weighted by the appropriate volume fractions of solid $(1-\phi)$ and fluid (ϕ):[11]

$$- \mathrm{tr}\underline{\underline{\sigma}}^m = - (1-\phi)\mathrm{tr}\underline{\underline{\sigma}} + 3\phi p . \qquad (35)$$

In (35) $\underline{\underline{\sigma}}$ is the average stress tensor within a grain, and 'tr' denotes the trace of a tensor.

The dependence of grain free energy on local stress follows from the stress dependence of the Gibb's free energy. The free energy U of a grain face in equilibrium with a medium exerting normal stress P_n takes the form[25]

$$U(P_n,\underline{\underline{\sigma}}) = \frac{P_n}{\rho} + \frac{F}{\rho_0} . \qquad (36)$$

ρ and ρ_0 are the molar densities of the solid in the stressed and reference states, respectively, and F is the Helmholtz or strain energy. We assume that the strain energy and molar density of the solid depend only on the average stress $\underline{\underline{\sigma}}$ in an isotropic, elastic solid and that the elastic deformation is infinitesimal and isothermal. This implies[8]

$$F = F(\underline{\underline{\sigma}}) , \quad \rho = \rho(\underline{\underline{\sigma}}) . \qquad (37)$$

In (37) we thus make the assumption that the average strain in the grain, as would be characteristic at the surface of grains at a distance away from grain contacts, is given by the average stress within the grain.

The average elastic strain energy in the solid is then taken in the form

$$F = \frac{1}{2} \, tr(\underset{=}{\varepsilon}\underset{=}{\sigma}) \quad . \tag{38}$$

F can be expressed solely in terms of two elastic constants and the mean and deviatoric parts of the average stress tensor:

$$tr(\underset{=}{\varepsilon}) = \frac{1}{3K} \, tr(\underset{=}{\sigma}) \, , \text{ and} \tag{39}$$

$$\hat{\underset{=}{\varepsilon}} = \frac{1}{2\mu} \, \hat{\underset{=}{\sigma}} \quad .$$

The diagonal components of the average stress tensor are found from the macroscopic stress tensor by expressions similar to equation (35), i.e.

$$\phi p \delta_{ij} - (1-\phi)\sigma_{ij} = -\sigma_{ij}^{m} \, , \quad i,j = 1, \, 2, \, 3 \tag{40}$$

such that by putting j=i, summing over i and dividing by three we regain (35). The components of the deviatoric part of the tensor are found from

$$\hat{\sigma}_{ij} = \sigma_{ij} - \frac{1}{3} \, tr\underset{=}{\sigma}\delta_{ij}, \tag{41}$$

and the strain energy is expressed as

$$F = \frac{1}{2}\left\{\frac{1}{9K}(tr\underset{=}{\sigma})^2 + \frac{1}{2\mu} \, tr(\hat{\underset{=}{\sigma}}^2)\right\} \quad . \tag{42}$$

Reaction at grain facets exposed to hydrostatically stressed pore fluid may be limited by interface detachment kinetics, or by diffusion through a leached layer. In either case, the rate of reaction at free faces, G_f, may be expressed as

$$G_f = -k^+ a\{1 - \exp(-A^f/nRT)\} \tag{43}$$

where k^+ is the dissolution rate coefficient (in cm/sec) at p and T, a is the activity of the solid (defined to be unity at conditions of p and T), A^f is the affinity for reaction at free faces, R is the gas constant, T is temperature, and n is a constant, close to unity. The coefficient k^+ depends on temperature, pore fluid chemistry, and other factors.

In the case of reaction at free grain facets, our assumption of an "averaged" strain energy yields the following expression for A^f:[8]

$$A^f = RT \left\{\ln K(p,T) - \sum_{\alpha=1}^{N} \nu_\alpha \ln a_\alpha\right\} + \gamma^f \, , \tag{44}$$

where

$$\gamma^f = \rho_o^{-1}\left\{p\frac{tr\underset{=}{\sigma}}{3K} + \frac{p^2}{K} + F - \frac{p^2}{2K}\right\} RT\ln a \equiv \gamma^f \tag{45}$$

$$\rho = \rho_o\{1-tr\underset{=}{\varepsilon}\}^{-1} \tag{46}$$

Stress-induced dissolution at grain contacts is thought to proceed by one of two mechanisms. In the water-film diffusion (WFD) mechanism, reaction occurs through a thin aqueous fluid film which supports the stress across the contact.[20] Reaction is limited by diffusion along the film and is driven by the normal stress dependence of Gibb's free energy. In the free-face pressure solution (FFPS) mechanism the normal stress applied to the solid is assumed to be the fluid pressure and hence, reaction is driven by elastic or permanent (dislocation) strain energies or possibly surface energy effects accompanying micro-fracturing at grain contacts, and may be limited by detachment/attachment kinetics or by diffusion.[26] It is likely that (nominally) monomineralic sandstones whose grain contacts have not experienced much shear displacement, are relatively clean, and have experienced relatively shallow burial will react by the FFPS mechanism, while sandstones containing clay or whose surfaces have experienced deformation conditions which act to increase the area of surface asperities in contact will react by the WFD mechanism. We will focus on the FFPS mechanism in this paper.

FFPS rates describing dissolution at asperities within grain contacts can be written in a form similar to (43). Following the suggestion of Ref. 26 that asperity contact areas will reach a steady state limited by dissolution (and not by plastic deformation or other mechanisms of asperity creep), we can describe grain shortening attending this type of pressure solution in terms of the dissolution rate. As shown in Ref. 8, for FFPS rates governed by elastic strain

$$G_z = k_z \left\{ c - \bar{c} \exp\frac{T(P_z-p)^2}{2E\rho_0 RT} \right\} , \qquad (47)$$

where

$$k_z = \left\{ \frac{\bar{c}}{k^+} + \frac{A_z \rho}{2D^c \Delta} \right\}^{-1} \qquad (48)$$

and T is the ratio of nominal to true contact area. As written, (47) reflects a diagonal stress tensor applied to contact surfaces with elements P_z, p and p; k_z captures both limits of diffusion and interface reaction controlled kinetics.

Inelastic Grain Deformation. As a grain dissolves at free faces, the contact area A_z can decrease. Hence, for large σ_{zz}^m without counterbalancing fluid pressure the stress across that contact can be very large. If this stress exceeds a critical value then the grains may deform inelastically. Because this deformation does not involve a change in grain volume (except for small and negligible elastic deformation or small volume increases accompanying microfracture) there is a direct relation between the change in L_z and L_f due to such an inelastic grain deformation. The critical vertical contact area below which the difference between P_z and p would exceed the compressive strength S_c is given by

$$A_{crit} = (-\sigma_{zz} - p)L_xL_y/S_c . \qquad (49)$$

If A_z falls below A_{crit}, due to free-face dissolution, plastic deformation driven by the difference in effective normal stress and compressive strength of quartz would act to increase the contact area to A_{crit}. To account for such plastic deformation, the rate of L_z due to an inelastic strain rate $\dot{\varepsilon}_{zz}^{(in)}$ is taken to be

$$\frac{\partial L_z^{(in)}}{\partial t} = L_z \dot{\varepsilon}_{zz}^{(in)} = \left\{ \begin{array}{ll} \hat{H}_z, & A_z < A_{crit} \\ 0, & A_z \geq A_{crit} \end{array} \right\} \equiv H_z \qquad (50)$$

In the case of a linear "visco-plastic" response, the contact maintains its rigidity until it decreases, by free-face dissolution, beyond a critical value A_{crit}, whereafter it "yields" at a rate proportional to the differential stress across the contact. $\hat{H}_z$ is given by

$$\hat{H}_z = \frac{1}{\eta} \left((P_z - p) - S_c \right) . \qquad (51)$$

The S_c term in (51) is subtracted from the differential stress to ensure that the deformation ceases as (P_z-p) approaches S_c. (P_z-p), and thus $\hat{H}_z$, decreases as the plastic deformation proceeds, so that the contacts in this case are "work hardening". We assume that this deformation is volume conservative (this would not be the case for micro-particle production due to granulation at contacts). To preserve a constant grain volume, the "mass" lost by a change in L_z must be accounted for by a corresponding increase in L_f. We add to $\hat{G}_f$ in in equation (3) a rate of change of grain radius H_f where

$$H_f = \frac{A_z}{A_f} H_z \; . \tag{52}$$

Summary of the Model. The mechanical-reaction-transport model has the following features:

* the microscopic mechanical problem is formulated in terms of Weyl's simple textural model;
* the textural variables L_z and L_f evolve by first-order time differential equations that generate local texture dynamics via a streaming term and a net mechanical deformation/reaction rate;
* the concentrations are obtained by the solution of modified, steady state reaction-transport equations coupled to algebraic conditions arising from the fast reactions in the pore fluid;
* the solute mass conservation equations are quasi-linear with respect to the concentration deviations γ_α due to the smallness of the driving force for the mechano-chemical dynamics;
* the mechanical problem is solved on the macroscopic level via effective porous medium theory with elastic coefficients modified by texture and a uniaxial "strain rate" due to the time dependent processes of pressure solution and inelastic grain deformation;
* the resulting mixed pde-type and algebraic constraint problem is nonlinear and highly coupled.

NUMERICAL SOLUTION. The variety of equation types and their strong coupling requires numerical techniques for solution. The system of equations is solved simultaneously by combining the ELLPACK system software[28] for elliptic equations (solute transport and elastic displacements) with a simple central-difference in space/forward difference in time algorithm for equations of rock texture. The ELLPACK subroutines for solving elliptical equations by banded gaussian elimination are accessed within an iterative procedure to yield stresses and pore fluid chemistry for a given rock texture. The textural variables L_z and L_f are then updated to the next time consistent with these. The rock flow velocity is found by applying trapezoidal summation to (5).

One-Dimensional Reactive Constraint on Horizontal Stresses. The horizontal components of the macroscopic stress tensor, and from them the horizontal derivatives of w_x and w_y, are found from a lateral constraint on the viscous deformation. A medium in which the compaction mechanism is predominantly by chemical means will cease compaction along a given direction when reaction at grain facets normal to this direction ceases. The normal components of the grain stresses, P_x and P_y, are thus determined by setting the horizontal growth rates G_x and G_y equal to zero (that is, setting G_x and G_y equal to 0 using analogous expressions to that given for G_z in (47)). P_x and P_y are found from the local texture and macro-stress components σ^m_{yy} and σ^m_{xx} from relations similar to (34). The simplified mechanical model is summarized as

$$\sigma_{zz}^m = \lambda^m \frac{\partial w_x}{\partial x} + \lambda^m \frac{\partial w_y}{\partial y} + (\lambda^m + 2\mu^m)\frac{\partial w_z}{\partial z} + \alpha^m p$$

$$\sigma_{xx}^m = \sigma_{yy}^m = -\left((P_y - p)\frac{A_y}{L_x L_y} + p\right)$$

$$\frac{\partial w_x}{\partial x} = \frac{\partial w_y}{\partial y} = \left(\frac{\sigma_{yy} - \lambda^m \frac{\partial w_z}{\partial z} - \alpha^m p}{2(\lambda^m + \mu^m)}\right)$$

(53)

$$\frac{\partial}{\partial z}\left((\lambda^m + \mu^m)\frac{\partial w_z}{\partial z} + \lambda^m \frac{\partial w_x}{\partial x} + \lambda^m \frac{\partial w_y}{\partial y} -+ \alpha^m p\right) = 0 \ .$$

Coordinate Transformations. In order to account for shortening arising from compaction in our finite-difference scheme, while maintaining a constant number of grid points, we introduce a new scaled spatial variable as follows. Let z denote the "real" spatial variable, and let ξ be the transformed quantity, ranging from 0 to 1, defined by

$$\xi = \frac{z - Z_0}{Z_b - Z_0} \ .$$

(54)

A change in z from Z_0 to Z_b (top to bottom) corresponds to a change in ξ from 0 to 1. To transform the evolution equations, note that for any variable ψ changing as a function of z and t ($\psi(z,t)$), there exists its corresponding transformed variable $\Psi(\xi,t)$. Partial derivatives with respect to both variables are related via:

$$\frac{\partial \psi}{\partial t} = \frac{\partial \Psi}{\partial t} - \frac{\partial \Psi}{\partial \xi}\left(\dot{Z}_0 + \xi(\dot{Z}_b - \dot{Z}_0)\right)/(Z_b - Z_0)$$

(55)

(the dots over Z_0 and Z_b denote differentiation with respect to time), and

$$\frac{\partial \psi}{\partial z} = \frac{\partial \Psi}{\partial \xi}(Z_b - Z_0)^{-1} \ .$$

(56)

By setting the flow velocity at the upper boundary to be equal to the subsidence velocity, u_0, the position of that boundary, Z_0, is given by

$$Z_0 = Z_0(t=0) + u_0 t$$

(57)

while the position of the lower boundary, Z_b, is given by

$$Z_b = Z_b(t=0) + \int_0^t u_z(Z_b)dt$$

(58)

The total length of the system is given by $Z = Z_b - Z_0$.

Boundary and Initial Conditions. The spatial grid upon which the equations are solved is seen as a one-dimensional sinking subvolume of rock on the sub-meter scale, oriented vertically with respect to gravity. The normal component of the elastic displacement, w_z, at the upper boundary is taken to be consistent with the lithostatic stress arising from the weight of the overlying water-saturated sediment. No-flux conditions for both solute and textural variables are imposed at both boundaries. Fluid pressure is taken to follow from the normal hydrostatic pressure gradient of .1 bar/meter; temperature is assumed to increase linearly with depth. We have neglected gradients in temperature, fluid pressure, and in stress due to a gravitational body force within the simulation domain. We point out that, although the boundaries are considered closed with respect to solute transport, they remain open with respect to pore-fluid transport. This ensures that pore-pressure remains at the hydrostatic level characteristic of the burial depth.

Simulation Results. We restrict our discussion here to simulations of monomineralic sandstones composed solely of quartz. For pore fluid pH at or below neutral, we may write the quartz-water reaction as

$$\text{quartz} \rightleftarrows SiO_2(aq) \qquad\qquad (59)$$

where $SiO_2(aq)$ denotes the solute component of silica in solution. This simple system is described in our model by five coupled partial differential equations – first order equations for the textural variables L_z and L_f and for the compaction velocity and second order equations for the deviatoric $SiO_2(aq)$ concentration and the vertical elastic displacement w_z.

Data for this system was chosen as follows: k^+, $\overline{c}$, ρ and elastic moduli are found in the literature.[29,30,31,32] The product $D^c\Delta$ is thought to lie in between 10^{-10} and 10^{-12} cm^3/s.[26] Other parameter values are discussed in the appropriate subsections to follow.

Spatial Self-Organization of Compaction and Cementation in Sandstones. To examine the consequences of spatial heterogeneity on the rates of chemical compaction and cementation, in this section we demonstrate a degree of "differentiation" between compaction and cementation in systems initialized with small-amplitude textural variations in L_z and L_f. These correspond to small grain-size variations in space; the wavelengths of variations are on the order of ones observed in depositional settings.[33]

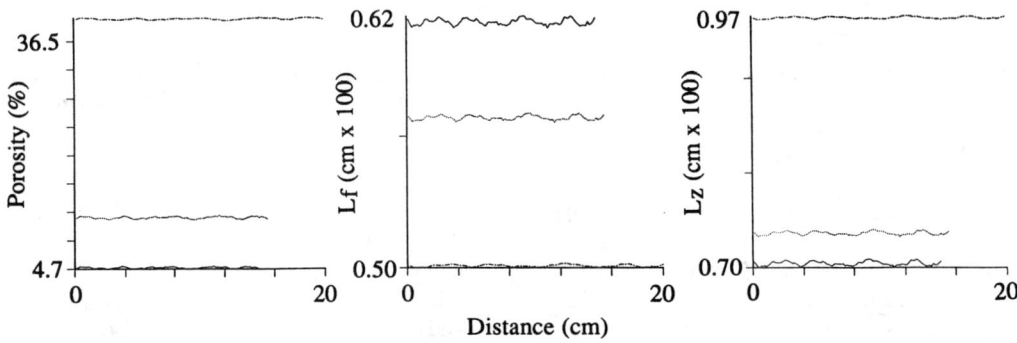

Figure 2. Evolution of small grain size variations in a very fine grained sandstone buried at 2.5 km. Shown are the porosity, and radius (L_f) and height (L_z) of grains at 0 (broken line), 40 (dotted line), and 80 million years (solid line). The simulation domain is oriented vertically with respect to the Earth's gravitational field, and shortens with time due to compaction. $D^c\Delta$ in this case is 10^{-10} cm^2/s; very little change in the amplitudes of the textural variations is evident at the final simulation time.

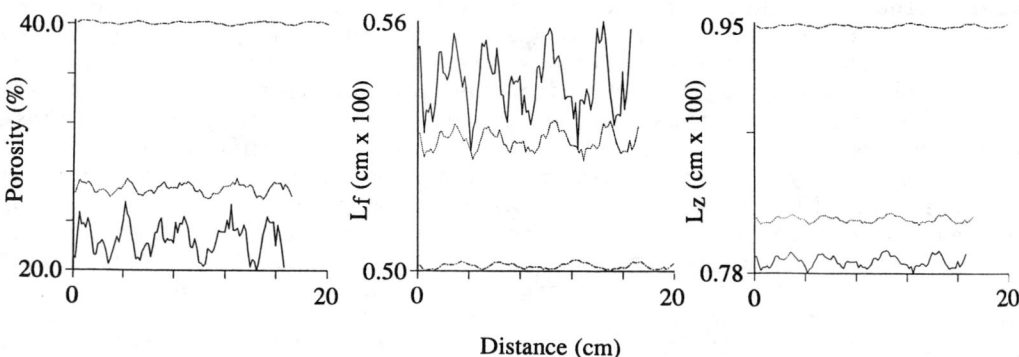

Figure 3. Same as in Figure 2, except here $D^c\Delta$ is 10^{-13} cm^3/s. As a result of the slower contact reaction rate, the system, undergoes a lesser amount of compaction than in Figure 2, but the amplitude of the spatial variations in texture increases with time. The zones of greatest compaction (lesser L_z) and porosity alternate with zones of greatest cementation.

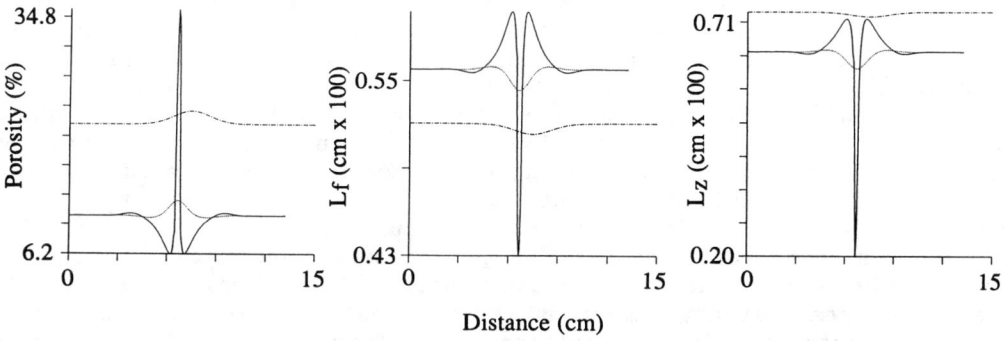

Figure 4. Development of a zone of highly localized dissolution and compaction by coupled free-face dissolution, contact dissolution, and plastic deformation. Shown are the porosity and radius (L_f), and height (L_z) of grains at 0 (broken line), 86.5 (dotted line) and 173 million years (solid line) for burial at 3 km. The development of the porosity maximum is likened to the evolution of a "joint", or plane of lost coherency, by stylolitization in sandstones.

We show in Figures 2 and 3 two simulations initialized with the same textural data but differing in the relative rates of free face and contact reaction. Figure 2 shows porosity, L_f and L_z profiles at run times 0 (broken line), 40 (dotted line), and 80 (solid line) million years. The amplitudes of the variations increase slightly with time. Figure 3, on the other hand, shows a larger amplitude pattern evolving, and a development in the spatial correlation of porosity, cementation (measured by increases in L_f) and compaction (compaction increases with decreasing L_z). The zones of greatest compaction alternate with those of the greatest cementation, and develop within those zones characterized by initial grain size minimums. The fluctuation of the textural variables from their spatial average increase as a function of decreasing the ratio of the reaction rate at contacts to that at free faces. The development of this style of spatial "differentiation" between cementation and compaction has been observed in sandstones[3,13,14] and in marl-limestone sequences.[1] The retention of greater porosity within zones of greater compaction in fact is characteristic of marl-limestone sequences.[1,34] The similarity between the styles of dissolution and cementation in sandstones compared to limestones suggests a degree of universality in the dynamics governing reaction and mass transport in these chemically dissimilar systems.

Stylolitization in Sandstones. Our mechanism for stylolitization in monomineralic rocks[8] involves free face dissolution, contact dissolution, and possibly plastic deformation of grains within stylolite zones. It appears to involve a "stronger" feedback mechanism than in the previous example. In Figure 4, we plot the textural variables of porosity, L_f and L_z for three simulation times of 0, 90, and 180 million years for a very fine-grained sandstone buried at 3 km. With time, the initial small amplitude "bump" in texture grows in a manner reminiscent of Figure 3, however, here dissolution and compaction become quite localized. The porosity in the stylolite zone increases because of the attending free face dissolution (we account for shortening only in the vertical direction, along the simulation domain, so that, with great amounts of free face dissolution, the vertical contact areas go to zero. Thus while compaction may be quite advanced in the vertical direction, the horizontal separation between grains remains constant). The enhanced cementation adjacent to the dissolution zone evident in the L_f profile by the final time is a common observation in naturally occurring stylolites.[1,3] Our simulations show that compaction and cementation occur relatively pervasively throughout the domain until a "critical" porosity is reached, whereafter any remaining textural heterogeneity in the system becomes unstable to stylolitization. This behavior, in which stylolitization proceeds only after a degree of pervasive intergranular pressure solution, has been invoked in sandstones by Heald.[13] It suggests that the development of stylolites in sandstones may be quite rapid relative to the overall compactional loss of porosity by pressure solution once certain critical criteria have been met.

The single heterogeneity in Figure 4 amplifies, and has induced secondary stylolites approximately 3 cm away by the final time shown. In this way, a periodic array of stylolites can arise from a single "parent".

The increases in the L_z profile in the regions adjacent to the developing stylolite suggest growth of quartz at the contacts against a normal stress. However, as can be seen in the positioning of the stylolite/porosity maximum, the overall rock shortening is zero once stylolitization has set in - the localized compaction at the stylolite center is taken up by small increases in L_z in the space surrounding the stylolite. We have found that this feature is characteristic in simulations of very-fine grained sandstone stylolitization evolving by the FFPS (and not the WFD) mechanism;[8] it is interesting in light of the compactional shortening of sandstone beds, but is an artifact of the (assumed) reversibility of the rate law (47).

Oscillatory Compaction Within Stylolite Zones

In the preceeding example, the localized compaction at stylolite zones took place by a combination of free-face dissolution, contact dissolution, and plastic deformation at grain-grain contacts. Figure 5 shows the evolution of the porosity within the stylolite zone with time, and indicates that it undergoes a minimum prior to the development of runaway dissolution. Once the stylolitic feedback takes off, small amplitude oscillations in porosity due to a dissolution/plastic deformation cycle commence. A magnified view of the oscillations in Figure 5 is seen in Figure 6(a). Figure 6(b) shows that while the porosity is increasing, the area of horizontal contacts between grains within the stylolite are undergoing small oscillations around a constant value. Plastic deformation may or may not occur during stylolitization by our model;[8] observation of its occurrence within stylolite zones is discussed in ref. 35.

Conclusions. The study, and more specifically, the modeling of chemical compaction in porous media holds much promise for interesting non-linear phenomena. We have shown that a formalism consisting of coupled equations describing rock/fluid interaction, diffusional mass transport, and visco-elastic deformation can describe the development of differential compaction and cementation as well as localized compaction in the form of stylolites. We suggest that shortening within stylolitic zones may occur by an oscillatory cycle of dissolution and plastic deformation.

The modeling studies described herein demonstrate a tendency toward self-organization in rock texture, and outline in parameter space conditions that favor the development of one behavior over another. Perhaps by modeling and predicting the occurrence of self-organized structures, we may say much about the history of a rock in the subsurface. From the point of view of petroleum geology, the development of self-organized banded compaction/cementation alternations and stylolites is of great potential importance regarding the question of preserved porosity and permeability at depth, and hence of deep petroleum reservoirs. As problems involving coupled sets of nonlinear partial differential and algebraic equations, these mechano-chemical systems present themselves as an interesting new area in applied mathematics.

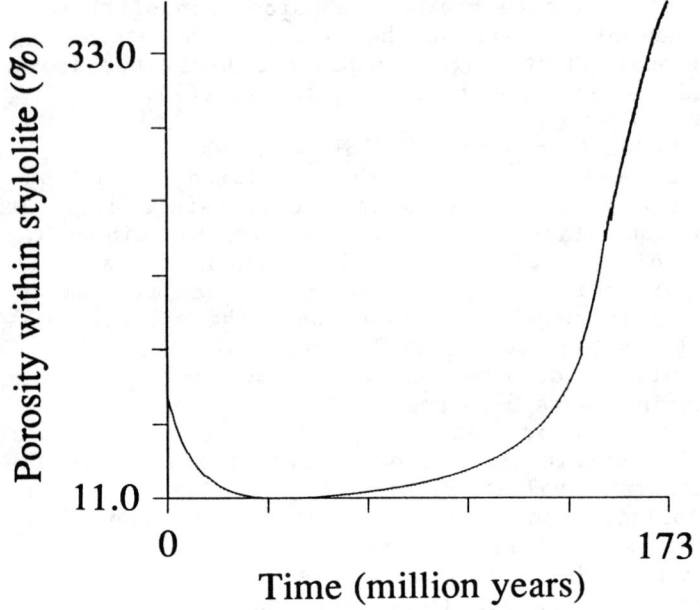

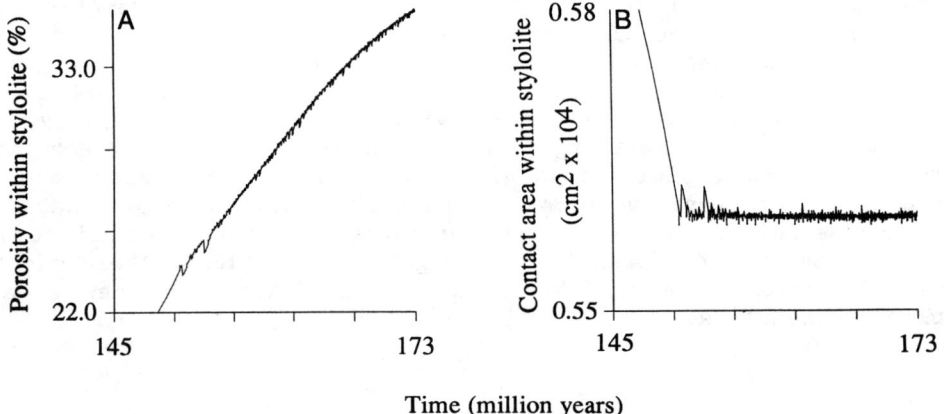

References

1. W. Ricken, Diagenetic Bedding, Springer Verlag, Berlin, 1986.

2. D. Houseknecht, Assessing the relative importance of compactional processes and cementation to the reduction of porosity in sandstones, AAPG Bull., 71 (1987) pp. 633-642.

3. R. Tada and R. Siever, Pressure solution during diagenesis: A review, Ann. Rev. Earth Planet. Sci., 17 (1989) pp. 89-118.

4. W. C. Park and E. H. Schot, Stylolites: Their nature and origin, J. Sed. Pet., 38 (1968) pp. 175-191.

5. R. A. Nelson, Localization of aggregate stylolites by rock properties, AAPG Bull., 67 (1983) pp. 313-322.

6. W. Alvarez, T. Engelder, and P. A. Geiser, Classification of solution cleavage in pelagic limestones, Geology, 6 (1978) 263.

7. E. Merino, P. Ortoleva, and P. Strickholm, Generation of evenly spaced pressure-solution seams during (late) diagenesis: A kinetic theory, Contrib. Min. Pet., 82 (1983) 360-370.

8. T. Dewers and P. Ortoleva, A coupled reaction/transport/mechanical model for intergranular pressure solution, stylolites, and differential compaction and cementation in clean sandstones, Geochim. Cosmochim. Acta, 54 (1990) pp. 1609-1625.

9. T. Engelder, A natural example of the simultaneous operation of free face dissolution and pressure solution, Geochim. Cosmochim. Acta, 46 (1982) pp. 69-74.

10. R. Tada and R. Siever, Experimental knife-edge pressure solution of halite, Geochim. Cosmochim. Acta, 50 (1986) pp. 29-36.

11. Carroll, M.M., Mechanical response of fluid-saturated porous materials, In Theoretical and Applied Mechanics; Rimrott, F.P.J.; B. Tabarrok, Eds.; North Holland Pub. Co., 1980, 251-266.

12. V. V. Palciauskas and P. A. Domenico, Fluid pressure in deforming porous rocks, Water Resources Research, 25 (1989) pp. 203-213.

13. M. T. Heald, Stylolites in sandstones, J. Geol., 63 (1955) pp. 101-114.

14. M. T. Heald and R. C. Anderegg, Differential cementation in the tuscarora sandstone, J. Sed. Pet., 30 (1960) pp. 568-577.

15. W. Ricken, The carbonate compaction law: A new tool Sedimentology, 34 (1987) pp. 571-584.

16. G. Rittenhouse, Pore-space reduction by solution and cementation, AAPG Bull., 55 (1971) pp. 80-91.

17. L. W. Teufel and J. M. Logan, Effect of displacement rate on the real area of contact and temperatures generated during frictional sliding of Tennessee Sandstone, Pure and Appl. Geophys. 1978, 116, 840–865.

18. P. Ortoleva, Geochemical Self-Organization, Oxford University Press, in press, 1992.

19. P. Ortoleva, Solute reaction mediated precipitate patterns in cross gradient free systems, Zeit. fur Fisik, B49 (1982) 149.

20. P. K. Weyl, Pressure solution and the force of crystallization, J. Geophys. Res., 64 (1969) 2001.

21. P. Ortoleva, E. Merino, J. Chadam and C. H. Moore, Geochemical self-organization I: Reaction-transport feedbacks and modeling approach, Am. J. Sci., 287 (1987) pp. 979–1007.

22. M. A. Biot and D. G. Willis, The elastic coefficients of the theory of consolidation, ASME Jour. Appl. Mechanics, 24 (1957) pp. 594–601.

23. Nur, A.; Byerlee, J.D., An exact effective stress law for elastic deformation of rock with fluids, Jour. Geophys. Res. 1971, 76, 6414–6419.

24. J. C. Jaeger and N. G. W. Cook, Fundamentals of Rock Mechanics, Chapman and Hall, London, 1979.

25. W. B. Kamb, The theory of preferred crystal orientation developed by crystallization under stress, Jour. Geol., 67 (1959) pp. 153–160.

26. R. Tada, R. Maliva and R. Siever, A new mechanism ,for pressure solution in porous quartzose sandstone, Geochim. Cosmochim. Acta, 51 (1987) pp. 2295–2301.

27. T. Dewers and P. Ortoleva, Geochemical self-organization III: A mean field, pressure solution model of spaced cleavage and metamorphic segregational layering, Am. J. Sci., 290 (1990) pp. 473–521.

28. J. R. Rice and R. F. Boisvert, Solving Elliptic Problems Using ELLPACK, Springer Verlag, New York, 1985.

29. J. D. Rimstidt and H. L. Barnes, The kinetics of silica-water reactions, Geochim. Cosmochim. Acta, 44 (1980) pp. 1683–1699.

30. R. O. Fournier, R.O. and R. W. Potter II, An equation correlating the solubility of quartz in water from 25ρ to 900ρC at pressures up to 10,000 bars, Geochim. Cosmochim. Acta, 46 (1982) pp. 1969–1973.

31. R. A. Robie, B. S. Hemingway and J. R. Fisher, Thermodynamic properties of minerals and related substances at 298.15 °K and 1 bar (10^5 Pascals) pressure and at higher temperatures, U.S. Geol. Survey Bull. 1452, 1979.

32. O. L. Anderson, E. Schreiber, R. C. Liebermann, and N. Soga, Some elastic constant data on minerals relevant to geophysics, Rev. Geophys., 6 (1968) 491–524.

33. J. T. Grace, B. T. Grothaus and R. Ehrlich, Size frequency distributions taken from within sand lamilae, Jour. Sediment Petrol. 1978, 48, 1193-1202.

34. T. Dewers and P. Ortoleva, Formation of stylolites, marl/limestone alternations, and clay seams through unstable chemical compaction of argillaceous carbonates, in Diagenesis, K. H. Wolf and G. V. Chilingarian, eds., in press.

35. C. J. R. Braithwaite, Mechanically-induced stylolites and loss of porosity in dolomites, Jour. Petroleum Geol. 1986, 9, 343-348.

CHAPTER 7

Conservation Laws that Change Type and Porous
Medium Flow: A Review**

Barbara Lee Keyfitz*

ABSTRACT: Systems of conservation laws
describing time evolution of a physical system
are expected, on the basis of linear theory,
to be of hyperbolic type. However, certain
complex flows, such as multiphase flow in
porous media, dynamic phase transitions in
elasticity, and flows with liquid-vapor
transitions, exhibit a change of type:
linearization about some states leads to

---

* Department of Mathematics, University of
Houston, Houston, Texas 77204-3476.
**This
paper includes a report on research supported by the AFOSR under grant AFOSR 86-0088,
and research supported jointly by NSF and
AFOSR under grant DMS-89-03768. Support from
the Texas Advanced Research Program, Project
Number 003652-077, is also acknowledged.

equations that are elliptic in space-time.

This paper is an exposition of some mathematical properties of such systems, including solution of the Riemann problem, shock admissibility criteria, regularization by higher-order terms, and numerical simulation. We also discuss the relation of systems which arise from modelling other complex flows to the examples which appear in modelling three-phase flow in porous media.

INTRODUCTION

At the SIAM/SEG/SPE meeting in Houston, January 1985, I attended the invited address given by Don Peaceman and had the pleasure of learning that some mathematical questions which I (among others) had been pursuing rather vaguely and abstractly were related to some real difficulties which had been observed in reactor simulation codes: it appeared, according to a careful analysis performed by Bell, Trangenstein and Shubin [3], that in modelling convection driven three-phase flow in a homogeneous porous medium - using the standard simplifying assumptions of Darcy's law and neglect of gravity and dispersion, and using a standard model, Stone's model, for the dependence of the three-phase relative permeabilities on the phase saturations - that one might end up with a system of conservation laws that was not hyperbolic in type. The equations, as Peaceman presented them (see [1] and [29]), and as Bell, Trangenstein and Shubin studied them, are

$$(1a) \qquad \frac{\partial s_w}{\partial t} + \frac{\partial f_w}{\partial x} = 0$$

$$(1b) \qquad \frac{\partial s_g}{\partial t} + \frac{\partial f_g}{\partial x} = 0$$

which express conservation of the aqueous

phase, s_w, and the gaseous phase, s_g, respectively. The relative saturation of the oleic phase is recovered from the equation

$$s_w + s_g + s_o = 1,$$

while all the complicated modelling assumptions are hidden in the description of the flux functions:

$$f_i(s_w, s_g) = \lambda_i / D, \quad i = w, g$$

where $\lambda_i = k_{ri}/\mu_i$ (the ratio of the relative permeability of the ith phase to its viscosity), and $D = \lambda_w + \lambda_g + \lambda_o$. This part of the model follows from the use of Darcy's law for a one- dimensional flow; to close the system and express the k_i as functions of the s_j requires the use of some specific model for the dependence when three phases are present, and here Stone's model was used. (The details are given in [3].)

Now, system (1) fails to be hyperbolic in the following sense. Compute the two-by-two Jacobian matrix, which we shall call A for the rest of this paper, of the flux vector, f; of course A depends on the vector of saturations, s. Inside the physically interesting triangle of nonnegative saturations, there are some small regions where the eigenvalues of A are complex. Incidentally, the precise location, size, and even number of these regions depends on the choice of "empirical" constants in Stone's model. Some natural questions about whether all choices in Stone's model lead to change of type in A, and whether alternative models might behave differently, are discussed in a more recent paper by Shearer and Trangenstein [42]; at the moment, one cannot prove categorically that all models for three-phase permeabilities will produce "elliptic

regions", but there is nothing obviously
unphysical about the models that have them.
An important special case is worth mentioning,
because it will reappear later: by making the
assumption $k_{ri} = k_{ri}(s_i)$ for all three
phases (this is a standard modelling
assumption for the aqueous and gas phases, but
is completely wrong for oil), one produces a
system where there is only one "elliptic
region", and it degenerates to a single point:
the eigenvalues of A are not complex here,
but are real and equal (such a point is called
an <u>umbilic</u>, a term which comes from
singularity theory), so the system is
nonstrictly hyperbolic.

I had been interested in conservation
laws with various bizarre properties - failure
to be strictly hyperbolic, for example [20,21]
- some of which also cropped up in
applications (in the miscible displacement
problem, addition of a polymer as in standard
enhanced recovery processes routinely yields a
type of nonstrictly hyperbolic system [10]),
and had learned a bit about shock wave
structure for these problems; like everyone
else in the field, I also knew of the
canonical example of a system that changes
type - steady, inviscid transonic flow [5] -
but the three-phase flow problem was different
from transonic flow because it was an
evolution problem (that is, the independent
variables were space and time), and linear
theory applied to system (1) predicts the
exponential amplification in time of
high-frequency waves for any initial data in
an elliptic region. This well-known result is
called the Hadamard instability. However,
computations in full-blown reservoir
simulators did not show any evidence at all of
instability; nor did the numerical
experiments performed by the authors of [3],
who at that time were working at Exxon and
were rather bemused that no one had perceived

the dangerously ill-posed nature of these equations. In fact, the computations of Bell, Trangenstein and Shubin were good enough that one could see the structure of all the shocks, and could see that they were all of either a classical "transonic" type (that is, a transition from a state in the hyperbolic region to one in the elliptic region of state space) or they were classical hyperbolic shocks, or they were limiting cases of these classical shocks, of a type that Kranzer and I had used when solving nonstrictly hyperbolic problems in [21].

The fact that no really new types of shocks were needed to resolve an arbitrary discontinuity in the initial conditions (this special type of data constitutes the so-called Riemann problem) was a particularly interesting feature of the Bell, Trangenstein and Shubin calculation, because of the one other example of an evolution system that changes type which had been worked out at that time. This was the system

$$\text{(2a)} \qquad \frac{\partial u}{\partial t} - \frac{\partial v}{\partial x} = 0$$

$$\text{(2b)} \qquad \frac{\partial v}{\partial t} - \frac{\partial p}{\partial x} = 0 \ ,$$

in the case that p is a nonmonotone function of u – take $p = u^3 - u$ for example. The strictly hyperbolic case, when p' is positive, serves as a classical example in conservation law theory: the equations of isentropic gas dynamics in Lagrangian coordinates are of this form [45]. When p is not monotone, system (2) is nonhyperbolic for values of u between the roots of $p' = 0$. This form for p appears in at least two different contexts: as a model for gas dynamics when the gas is undergoing a phase transition of van der Waals type below the critical temperature [43], and as a model for one-dimensional two-phase elasticity [12].

More recently, system (2) has been proposed as a model for the dynamics of other two-phase mixtures [2,46].

Motivated by some conjectures in [12], Shearer worked out the solution of the Riemann problem for (2) when p has a roughly cubic form [37]. A useful way of thinking of this problem is to imagine bringing the two roots of p′ together: the limiting case is a hyperbolic system which fails to be strictly hyperbolic along a line in the uv plane. This degeneracy turns out to have quite different consequences from the umbilic degeneracy. In fact, a nonstrictly hyperbolic system of the form (2) is quite well behaved, as Kranzer and I had discovered in [21], and Shearer's solution, in [37], was very much like ours except for a strip in phase space in which the influence of the nonhyperbolic region could be seen. But in that strip, Shearer showed, there is no way of solving the Riemann problem using only classical shocks. Shearer introduced a new kind of shock, called <u>undercompressive</u>, which does not satisfy the classical Lax geometric entropy condition on the number of entering characteristics. Comparisons between the two systems, and stability results for solutions of the Riemann problem for (2), appear in [13,38,39].

So there are two striking facts: a system, (1), of a bizarre mathematical nature, appears in a genuine application, and turns out already to have attracted the attention of mathematicians; but the solution does not match what has been computed in prototype problems. This paradox has stimulated a lot of research in the last few years. Some attention has turned recently to basic mathematical questions - what is meant by well-posedness when the equations are linearly ill-posed in the classical sense - and some of it has focussed on the models themselves: are the solutions of systems (1) and (2) essentially different from each other, and, more generally, how can one classify systems that change type; and does system (1) or (2) really represent an appropriate limit of a physically more complete system? These questions have not been completely answered yet, but there is some progress, and I will try, in the rest of this paper, to describe what is now known about systems that change type. Before doing so, I will mention another example in which equations that change type seem to occur.

The context is unsteady flow of two distinct phases. For the simplest case [32], consider one-dimensional flow in a pipe of constant cross-section of two separated fluids, each ideal: compressible, inviscid, nonreacting, and isentropic, each with density given by a gas law $\rho_i(p)$, $i = 1,2$, where p is the pressure across the channel. Suppose u_i is the velocity and α_i the (linear) volume fraction of fluid i; ignore surface tension at the interface. Then the flow is governed by a system of four equations for u_1, u_2, α_1 and p (we can solve for $\alpha_2 = 1-\alpha_1$) as functions of x and t:

(3a)
$$\frac{\partial}{\partial t} (\alpha_1 \rho_1) + \frac{\partial}{\partial x} (\alpha_1 \rho_1 u_1) = 0$$

(3b)
$$\frac{\partial}{\partial t} (\alpha_2 \rho_2) + \frac{\partial}{\partial x} (\alpha_2 \rho_2 u_2) = 0$$

(3c)
$$\frac{\partial}{\partial t} (\alpha_1 \rho_1 u_1) + \frac{\partial}{\partial x} (\alpha_1 \rho_1 u_1^2) + \alpha_1 \frac{\partial p}{\partial x} = 0$$

(3d)
$$\frac{\partial}{\partial t} (\alpha_2 \rho_2 u_2) + \frac{\partial}{\partial x} (\alpha_2 \rho_2 u_2^2) + \alpha_2 \frac{\partial p}{\partial x} = 0 .$$

A calculation reveals that this system also changes type: two characteristic speeds are always real and the other two behave like the roots of (1): they are real or complex depending on the states. Although this problem may seem rather special, it has been much studied because if, instead of assuming that the fluids are separated, one supposes they are mixed and one averages all quantities across the channel, essentially the same equations are found. This is discussed in the review article by Stewart and Wendroff [47], who also explain the importance of this model in nuclear reactor design, where the fluids are water and steam in cooling pipes.

The nonhyperbolic nature of this system, and attendant difficulties [6], seem to be somewhat better known than is the porous medium problem now becoming familiar to applied mathematicians who look at reservoir simulation problems, and a number of attempts to fix it are documented. They all follow the route of assuming that some essential physics has been omitted from (3), and proceed by, for example, introducing surface tension [32], or by supposing that the pressure may be different in the two fluids and then closing the system by adding a transport equation [33]. Of course, it is never incorrect to add more physics. Ransom and Hicks, in [33], compare numerical simulations on (3), which contain oscillations large enough to destroy any accuracy, with simulations on a two-pressure model, where clean, monotone shocks are produced. It is tempting to suppose that the oscillations are evidence of the Hadamard instability, and that they disappear when the problem is rendered hyperbolic. However, this hypothesis has not been tested; in fact, I conjecture that the oscillations result from the use of a Lax-Wendroff scheme, and that the first-order scheme of [3] would compute a reasonable solution for the mixed-type problem (3).

These examples present typical problems which face an applied mathematician: what can we say about a new kind of equation which appears when we apply well-known modelling

principles to a slightly unfamiliar situation?
How do we know when a given model contains
enough physics to model phenomena of interest
with sufficient accuracy for the situation at
hand? And, in the absence of a complete
understanding of the mathematical properties
of the solution, how do we know when a chosen
numerical scheme is correctly simulating the
desired solution? From the mathematical
viewpoint, what the examples I have described
have in common is that they all generalize
textbook derivations, and yet produce
nonhyperbolic systems. The results I shall
describe in the next sections do not address
directly the question of whether these models
include sufficient physics or not (this
remains an open question; it is discussed in
some of the references). We examine first the
mathematical question of how nonlinear
properties of conservation laws may influence
nonhyperbolic systems and may mitigate the
ill-posedness of the latter.

THE LOCAL STRUCTURE OF SYSTEMS THAT CHANGE TYPE

Consider a system of two conservation
laws involving a $\underline{state}$ $\underline{variable}$ $w = (u, v) = (w_1, w_2)$:

$$(4) \qquad \partial_x \begin{pmatrix} f_1(w) \\ f_2(w) \end{pmatrix} + \partial_y \begin{pmatrix} g_1(w) \\ g_2(w) \end{pmatrix} = 0 .$$

Here f and g are, in general, nonlinear
$\underline{flux}$ $\underline{vectors}$. Systems (1) and (2) are
examples of (4) with $y = t$ representing the
time variable, and $g(w) = w$. System (3) does
not fit into this framework because it is not
in conservation form and there are four
equations. As mentioned in the previous
section, the change of type in (3) is
restricted to two characteristic variables,
and, as is proved in [19], (3) can be reduced

locally to a system of two equations, which
are in conservative form. (The
nonconservative form of (3) is inconsistent
with the theory outlined in this paper; some
researchers recommend replacing (3) with a
system in conservation form, which can be done
[19] and [31]; but see [4] and [23] for
another approach.) System (4) can also be
specialized to steady inviscid transonic
potential flow in two dimensions:

(5a) $$\frac{\partial}{\partial x} (\rho u) + \frac{\partial}{\partial y} (\rho v) = 0$$

(5b) $$\frac{\partial v}{\partial x} - \frac{\partial u}{\partial y} = 0$$

where x and y are spatial coordinates and
$\rho = \rho(|w|)$ is the density, given by
Bernoulli's equation.

We write system (4) in quasilinear form:

(6) $$P(w, \nabla w) = A(w) \frac{\partial w}{\partial x} + B(w) \frac{\partial w}{\partial y} = 0 ,$$

where $A = df$ and $B = dg$ are the Jacobian
matrices of the fluxes. Now, (4) or (6) is
<u>hyperbolic</u> at a state w if

(7) $$\det | A \xi + B \eta | = 0$$

has nontrivial real solutions (ξ, η); if B
is invertible, then this is equivalent to
saying that the eigenvalues λ of $-B^{-1}A$
are real. Since (7) is homogeneous, it is
sufficient to consider dual vectors of unit
length: $(\xi, \eta)' = (\cos \beta, \sin \beta)$. Let $\mathcal{H}$ be
the set of hyperbolic states: for $w \in \mathcal{H}$,
there are two real roots, $\beta_1 < \beta_2$; in the

generic local picture for change of type
[17,28], $\mathcal{H}$ is an open set whose boundary,
$\mathcal{B}$, is a smooth curve, called the <u>sonic line</u>
in transonic flow. The complement, $\mathcal{E}$,
consists of <u>nonhyperbolic</u> states; in the case
$w \in \mathbb{R}^2$, equation (4) is <u>elliptic</u> at such
states.

The important consequence of hyperbolicity for physical modelling is often expressed in the statement that the characteristic speeds, λ, are real. But a more complete description, valid also in the case that both x and y represent spatial coordinates, is to say that the <u>domain</u> <u>of</u> <u>dependence</u> of the linearized equation

(8) $P(w_0, \nabla w) = 0$

is a proper subset of the xy plane precisely for hyperbolic points w_0. A careful

definition of this requires the notion of a <u>fundamental</u> <u>solution</u>: there exists a solution of (8) which is nonzero in exactly one quadrant of the xy plane, where quadrants are separated by the <u>characteristic</u> <u>surfaces</u>

(9) $\phi(x, y) = x \cos \beta + y \sin \beta = 0$

which are normal to the characteristic covectors determined by β_1 and β_2. If B is invertible, these are

$$\frac{dx}{dy} = \lambda_i ,$$

and, in the familiar case that y represents time, the distinguished quadrant, or <u>support</u> <u>cone</u>, is

$$\lambda_1 y \leq x \leq \lambda_2 y .$$

This elementary point is important: when we look at nonlinear theory, the choice of the forward time cone, from the four quadrants that are theoretically possible, is equivalent to postulating irreversibility of the system and hence to a determination of the admissible weak solutions, at least as far as they are determined by conditions of admissibility of shocks [25]. When y is not a time variable, as in steady transonic flow, a statement that the domain of dependence is finite is not equivalent to finite speed of propagation, but instead has as a consequence that influence propagates only downstream in the supersonic region. In this case the mathematically arbitrary but physically significant choice of support cone is the quadrant which contains the flow vector, and, with this choice, the standard admissibility conditions for fully supersonic shocks are the same as for the unsteady problem. (See [18] and [22].)

There is one important difference between the standard examples of unsteady flow as in equation (1) and steady flow of equation (5): as w_0 approaches $\mathcal{B}$, the sonic line, the support cone in the unsteady case narrows to a singular line; in the steady case, it widens to an open half-plane [18]. Intuitively, one might think of the distinction as follows: supersonic steady flow near the sonic line is almost like subsonic steady flow, in which a perturbation at a point influences the entire flow field; by contrast, unsteady systems with nearly coincident characteristic speeds act resonantly, with an effect concentrated in the characteristic direction. The term resonance has been introduced recently by Liu to describe properties of nonstrictly hyperbolic systems [24].

To explain a connection between hyperbolicity and weak solutions of conservation laws, let me recall the definition of a <u>shock solution</u> of (5). Locally, a shock has the structure

$$(11) \qquad w(x, y) = \begin{cases} w_0, & x \cos \beta + y \sin \beta < 0 \\ w_1, & x \cos \beta + y \sin \beta > 0 \end{cases}$$

where w_0, w_1 and β are related by the <u>Rankine-Hugoniot</u> condition:

$$(12) \qquad V(w_1, w_0, \beta) \equiv (f(w_1) - f(w_0)) \cos \beta \\ + (g(w_1) - g(w_0)) \sin \beta = 0$$

and now β is not a characteristic angle if
system (4) is genuinely nonlinear. Now, it is
well known that not all discontinuous
solutions like (11) correspond to "physical"
states of system (5): for example, if (11) is
admissible, then its mirror image, with w_0

and w_1 interchanged, is also a weak

solution, and is generally not admissible.
This interchange corresponds, roughly, to
reversing time, and admissibility can be
described in terms of dissipating energy or
increasing entropy, as well as in other
mathematical or physical ways [45]. Thus, the
notion of admissibility is intrinsically
related to hyperbolicity through the choice of
support cone described above.

A curve is <u>spacelike</u> if its tangent is
everywhere exterior to the support cone, and
for the standard choices of support cone in
both unsteady hyperbolic and steady transonic
systems, an admissible shock is spacelike with
respect to precisely one of the two states,
and this is the <u>upstream</u> state. This
coincides with the Lax geometric criterion
[25] for unsteady flow, and with the standard
requirement that shocks be compressive in
transonic flow [5]. Furthermore, a detailed
examination of the geometry, which was carried
out in [18], shows that unsteady systems which
change type, such as (1), are fundamentally
different from (5) at this level. I have
suggested elsewhere that some extension of
these ideas might give precise definitions of
stability for shocks in (1), but this has not
been carried out yet.

Before going on to outline other
explorations of shock admissibility, I should
mention one other result which shows how
shocks (or weak solutions) are basically a
hyperbolic phenomenon: when a solution like
(11) exists, it is necessary for at least one
of the states w_i to be in $\mathcal{H}$, at least as

long as $\mathcal{E}$ is convex and connected. This was proved by Holden, Holden and Risebro in [9]. Some other results on the shape of the Hugoniot locus for states near $\mathcal{B}$ can be found in [16] and [28].

Another approach to shock admissibility for (1) or other systems with a time variable is to perturb the system by including a higher-order dissipative term, converting (1), for example, to

$$(13) \qquad \frac{\partial s}{\partial t} + \frac{\partial f}{\partial x} = \frac{\partial}{\partial x} \left(D \frac{\partial s}{\partial x} \right)$$

where D is a positive definite or semidefinite matrix, or at least has eigenvalues of nonnegative real part. In the context of three-phase flow, D represents dispersive terms, and a typical form for D in this case is given in [3]. More generally, (13) is a parabolic regularization of (1), and the general system (4) can be likewise regularized, though if x and y are both spatial variables, the form of the higher-order terms depends on the chosen support cone (see [22] and [26]). Numerical schemes that approximate (1) are often considered to be approximations of (13) instead, and D is often called "artificial viscosity". A particularly straightforward choice for D is $D = I$; this has the property of preserving the invariant regions of the hyperbolic problem [45]. I have considered the relation between viscous perturbations and admissibility for shocks near the sonic line in [16] and [18], using bifurcation theory for equilibria and vectorfields. An extensive literature on viscous profiles for nonstrictly hyperbolic systems and systems with elliptic regions has arisen from modelling studies, and from well-posedness criteria for the Riemann problem. We turn to this now.

THE GLOBAL STRUCTURE OF ELLIPTIC REGIONS AND RIEMANN PROBLEMS

The arguments in the preceding section indicate that shocks joining points in the elliptic region to hyperbolic states may be admissible or stable in some sense. They suggest, and even quantify, a notion of wave speed for elliptic states by associating with a state in the elliptic region a nonlinear shock wave joining it to a hyperbolic state. Other mathematically oriented research in this subject has attacked more global questions: the geometry of the elliptic region itself and the effect of perturbing the region. Shearer and coworkers have made clever use of the following observation: to study the eigenvalues of a matrix A, we may replace it by its <u>deviator</u>, the trace-free matrix $A - (trA/2)I$. Writing this matrix in the form

$$\begin{pmatrix} a & b+c \\ b-c & -a \end{pmatrix},$$

we note that the map $w \mapsto (a, b, c)$ defines a surface in $\mathbb{R}^3$; the portions of the surface that are in the interior of the cone $c^2 = a^2 + b^2$ correspond to the elliptic regions. States that map to the vertex of the cone are umbilic points. If the boundary of state space maps to a loop that encircles the cone, then either an umbilic point or an elliptic region exists in state space; see Schaeffer and Shearer [34] and Shearer and Trangenstein [42]. In general, elliptic regions may be of two types: those that can be perturbed away (their boundary may be deformed to a point without passing through the vertex of the cone), and those that cannot. A special case that has been much studied is the case of quadratic flux functions: when f is a homogeneous quadratic, the origin is always a point of nonstrict hyperbolicity, and one can classify the quadratic fluxes for which the

system is strictly hyperbolic in the
complement. A few of the numerous studies of
these models and their very intricate Riemann
solutions may be found in the references
[11,34,35,40,41]. It is noteworthy that even
without the presence of elliptic regions these
Riemann problems have nonclassical solutions -
especially the so-called undercompressive
shocks mentioned earlier - and questions of
uniqueness and admissibility have not been
completely answered.

These models first arose in studying
local behavior of prototypes for the
three-phase saturation model (1).
Specifically, the assumption $k_{ri} = k_{ri}(s_i)$
leads to an umbilic point; and a Taylor
expansion to quadratic terms about this point,
followed by a perturbation by a linear term,
yields an elliptic region which is a
perturbation of the umbilic. Of course, the
same complicated phenomena will occur (and
possibly others); several cases have been
examined by Holden and others [7,8,9]. Here,
also, there are not yet definitive answers on
existence or uniqueness, even for Riemann
problems. But it is noteworthy that the
complicated features of these change of type
models are already present in the hyperbolic
prototype. Thus, the potential instabilities
are not merely the result of the presence of
elliptic regions.

Perturbed umbilic points are not the only
way elliptic regions can occur, however. In
[14] and [15], I suggested that a different
model might have more in common with the
calculations done in [3] on Stone's model. By
contrast with the quadratic models, the class
I considered is described by a singular
mapping: $w \mapsto (a, b, c)$ is a line, not a
surface, and crosses the boundary of the cone
far from the vertex. This model required many
assumptions - too many, I think, to make it
immediately applicable to saturation models.

A relation to quadratic models is explored in [17]; see also [11]. However, there is a common theme in these models, and in the change of type model (2), which may be important: they can all be derived by perturbation from a nonstrictly hyperbolic system where the coincidence of eigenvalues occurs along a line in phase space rather than at an umbilic point. Thus, the model for the elliptic region is a strip, not a cirle. It is still possible, as Shearer's study of (2) showed [37], that the waves will involve undercompressive shocks; in the example I studied [15], undercompressive shocks were avoided by making assumptions on the variation of the wave speeds on the two sides of the elliptic strip. But the geometry and the analysis in both cases are much simpler than for perturbations of an umbilic point. Furthermore, failure of strict hyperbolicity along a line occurs in a number of classical problems [20] and is not associated with ill-posedness. The matter of which analogy is closer to the three-phase model (1) is definitely worth pursuing.

We close this section with a comment on the invariance of the hyperbolic region: it was noted in [3] and also more generally in [30] and [9] that the solution of the Riemann problem, for initial states that are in $\mathcal{H}$, lies completely in the hyperbolic region. This seemed to suggest a paradox [1, p146]: some states in the model (1), while not physically or thermodynamically unstable, are "avoided" by the flow. However, Pego and Serre [30] and Holden et al. [9] have shown that this is not true for general Cauchy data: smooth solutions may well enter the elliptic region. This has further implications - for example, what happens to approximations which use Riemann problems, such as the Glimm scheme or Godunov schemes? An interesting analysis of one such approximation to a smooth flow is

given by Pego and Serre in [30].   Further
studies in this direction will be very useful.

CONCLUSIONS

The discovery - or rediscovery - of
change of type in simplified reservoir models
has stimulated much ongoing mathematical
research, numerical simulation, and even a
re-examination of basic mathematical
assumptions on what constitutes well-posedness
for quasilinear problems. At the moment, the
topic has a somewhat experimental flavor:
current research is branching out in several
directions and it would be hard to describe a
unified theory. In this exposition, I have
not given a technical summary of the many
partial results that are now available; for
these, the reader can consult the references.
I have tried to indicate some of the new
approaches. There will be more in the future.
It is too soon to say that these results will
be important, in any practical way, to
reservoir simulation. But this work has made
it clear that the skeptical questioning of
established meta-truths is healthy. Don
Peaceman began this trend with his exciting
talk of five years ago. You will undoubtedly
hear more at the next meeting.

REFERENCES

1.   M. B. Allen III, G. A. Behie, and J. A. Trangenstein, *Multiphase Flow in Porous Media,* Springer, New York, 1988.

2.   H. W. Alt, K.-H. Hoffman, M. Niezgódka, and J. Sprekels, "A numerical study of structural phase transitions in shape memory alloys", Preprint No. 90, Inst. for Math., University of Augsburg, 1985.

3.   J. B. Bell, J. A. Trangenstein and G. R. Shubin, "Conservation laws of mixed type describing three-phase flow in porous media", *SIAM Jour. Appl. Math.* 46 (1986), 1000-1023.

4.   J. J. Cauret, J. F. Colombeau and A. Y LeRoux, "Discontinuous generalized solutions of nonlinear nonconservative hyperbolic equations", *Jour. Math. Anal. Appl.* (in press).

5.   R. Courant and K. O. Friedrichs, *Supersonic Flow and Shock Waves*, Wiley Interscience, New York, 1948.

6.   D. Gidaspow, in "Suggestions for further

research and urgent problems on two-phase flows and heat transfer" by A. E. Bergles. et. al., *Two-Phase Flows and Heat Transfer Vol.III*, (ed. Kakaç & Veziroglu) Hemisphere Publishing Corp., Washington, 1976, 1457-1458.

7.   H. Holden, "On the Riemann problem for a prototype of a mixed type conservation law", *Comm. Pure Appl. Math.* 40 (1987), 229-264.

8.   H. Holden and L. Holden, "On the Riemann problem for a prototype of a mixed type conservation law, II", in *Current Progress in Hyperbolic Systems: Riemann Problems and Computations*, (W. B. Lindquist ed), AMS, Providence, 1989, 331-367.

9.   H. Holden, L. Holden and N. H. Risebro, "Some qualitative properties of 2x2 systems of conservation laws of mixed type", to appear in *Nonlinear Evolution Equations that Change Type* (ed. B. L. Keyfitz and M. Shearer).

10.   E. Isaacson, "Global solution of a Riemann problem for a non-strictly hyperbolic system of conservation laws arising in enhanced oil recovery", *Jour. Comp. Phys.*, to appear.

11.   E. Isaacson, D. Marchesin and B. Plohr, "Transitional shock waves", in *Current Progress in Hyperbolic Systems: Riemann Problems and Computations*, (W. B. Lindquist ed), AMS, Providence, 1989, 125-145.

12.   R. D. James, "The propagation of phase boundaries in elastic bars", *Arch. Rat. Mech. Anal.* 73 (1980), 125-158.

13.   B. L. Keyfitz, "The Riemann problem for nonmonotone stress-strain functions: a 'hysteresis' approach", in *Nonlinear Systems of Partial Differential Equations in Applied Mathematics*, (Basil Nicolaenko, ed), AMS, Providence, 1986, 379-395.

14.   B. L. Keyfitz, "An analytic model for

change of type in three-phase flow", in *Numerical Simulation in Oil Recovery*, (ed M. F. Wheeler), IMA Vol 11 (1988), Springer, New York, 149-160.

15.   B. L. Keyfitz, "Change of type in three-phase flow: a simple analogue", *Jour. Diff. Eqns.* 80 (1989) 280-305.

16.   B. L. Keyfitz, "The use of vectorfield dynamics in formulating admissibility conditions for shocks in conservation laws that change type", in *Problems Involving Change of Type* (ed. K. Kirchgassner), Springer, Berlin, 1990.

17.   B. L. Keyfitz, "A criterion for certain wave structures in systems that change type", in *Current Progress in Hyperbolic Systems: Riemann Problems and Computations*, (W. B. Lindquist ed), AMS, Providence, 1989, 203-213.

18.   B. L. Keyfitz, "Shocks near the sonic line: a comparison between steady and unsteady models for change of type", to appear in *Nonlinear Evolution Equations that Change Type* (ed. B. L. Keyfitz and M. Shearer).

19.   B. L. Keyfitz, "Change of type in simple models for two-phase flow", University of Houston Mathematics Department research report UH/MD-66, 1989, submitted to proceedings from the *Workshop on Viscous and Numerical Approximation of Shock Waves* (ed. Michael Shearer).

20.   B. L. Keyfitz and H. C. Kranzer, "A system of hyperbolicconservation laws arising in elasticity theory", *Arch. Rat. Mech. Anal.*, 72, (1980), 219-241.

21.   B. L. Keyfitz and H. C. Kranzer, "The Riemann problem for a class of hyperbolic conservation laws exhibiting a parabolic degeneracy", *Jour. Diff. Eqns.* 47 (1983), 35-65.

22.   B. L. Keyfitz and G. G. Warnecke, "The existence of viscous   profiles and admissibility for transonic shocks", preprint, submitted to *Commun. in Partial Diff. Eqns.*

23.   P. LeFloch, "Entropy weak solutions to nonlinear hyperbolic systems under nonconservative form", *Commun. in Partial Diff. Eqns.* 13 (1988), 669-727.

24.   T.-P. Liu, "Nonlinear resonance for quasilinear hyperbolic equations", *J. Math. Phys.* 28 (1987), 2593-2602.

25.   A. Majda, *Compressible Fluid Flow and Systems of Conservation Laws in Several Space Variables*, Springer, New York, 1984.

26.   A. Majda and R. L. Pego, "Stable viscosity matrices for systems of conservation laws", *Jour. Diff. Eqns.* 56 (1985), 229-262.

27.   D. Marchesin and H. B. Medeiros, "A note on the stability of eigenvalue degeneracy in nonlinear conservation laws of multiphase flow", in *Current Progress in Hyperbolic Systems: Riemann Problems and Computations*, (W. B. Lindquist ed), AMS, Providence, 1989, 215-224.

28.   M. S. Mock, "Systems of conservation laws of mixed type", *Jour. Diff. Eqns.* 37 (1980), 70-88.

29.   D. W. Peaceman, *Fundamentals of Numerical Reservoir Simulation*, Elsevier, Amsterdam, 1977.

30.   R. L. Pego and D. Serre, "Instabilities in Glimm's scheme for two systems of mixed type", *SIAM J. Numer. Anal.* 25 (1988), 965-988.

31.   B. J. Plohr and D. H. Sharp, "A conservative Eulerian formulation of the equations for elastic flow", *Adv. Appl. Math.* 9 (1988), 491-499.

32.   J. D. Ramshaw and J. A. Trapp, "Characteristics, stability, and short-wavelength phenomena in two-phase flow equation systems", *Nuclear Sci. and Eng.* 66 (1978), 93-102.

33.   V. H. Ransom and D. L. Hicks, "Hyperbolic two-pressure models for two-phase flow", *Jour. Comp. Phys.* 53 (1984), 124-151.

34.   D. G. Schaeffer and M. Shearer, "The classification of 2x2 systems of nonstrictly hyperbolic conservation laws, with application to oil recovery", *Comm. Pure Appl. Math.* 40 (1987), 141-178.

35.   D. G. Schaeffer and M. Shearer, "Riemann problems for nonstrictly hyperbolic 2x2 systems of conservation laws", *Trans. Amer. Math. Soc.* 304 (1987), 267-306.

36.   S. Schecter and M. Shearer, "Undercompressive shocks for nonstrictly hyperbolic conservation laws", 1989, preprint.

37.   M. Shearer, "The Riemann problem for a class of conservation laws of mixed type", *Jour. Diff. Eqns.* 46 (1982), 426-443.

38.   M. Shearer, "Admissibility criteria for shock wave solutions of a system of conservation laws of mixed type", *Proc. Royal Soc. Edinburgh* 93A (1983), 233-244.

39.   M. Shearer, "Nonuniqueness of admissible solutions of Riemann initial value problems for a system of conservation laws of mixed type", *Arch. Rat. Mech. Anal.* 93 (1986), 45-59.

40.   M. Shearer, "The Riemann problem for 2x2 systems of hyperbolic conservation laws with case I quadratic nonlinearities", *Jour. Diff. Eqns.* 80 (1989), 343-363.

41.   M. Shearer, D. G. Schaeffer, D. Marchesin and P. L. Paes-Leme, "Solution of the Riemann problem for a prototype 2x2 system of

non-strictly hyperbolic conservation laws",
Arch. Rat. Mech. Anal. 97 (1987), 299-320.

42.   M. Shearer and J. Trangenstein, "Change
of type in conservation laws for three phase
flow in porous media", *Transport in Porous
Media*, to appear.

43.   M. Slemrod, "Admissibility criteria for
propagating phase boundaries in a van der
Waals fluid", *Arch. Rat. Mech. Anal.* 81
(1983), 301-315.

44.   M. Slemrod, "A limiting 'viscosity'
approach to the Riemann problem for materials
exhibiting change of phase", *Arch. Rat. Mech.
Anal.* 105 (1989), 327-365.

45.   J. Smoller, *"Shock Waves and Reaction-
Diffusion Equations"*, Springer-Verlag, New
York, 1983.

46.   J. Sprekels, "Global existence for
thermomechanical processes with nonconvex free
energies of Ginzburg-Landau form", *Jour. Math.
Anal. Appl.*, to appear.

47.   H. B. Stewart and B. Wendroff, "Two-phase
flow: models and methods", *Jour. Comp. Phys.*
56 (1984), 363-409.

CHAPTER 8

Geology and Mathematics: Scientific Control of Art
and Legerdemain

I. Lerche*

Abstract.   A brief review is presented of the three
dominant, intertwined, threads: burial, thermal and
hydrocarbon histories. The rationale for requiring an
integrated picture of basinal evolution is examined as is
the use of control information at particular well sites to
constrain quantitative model behaviors. Input well data
in the form of present-day formation thicknesses, porosity
with depth, thermal gradient, vitrinite reflectance with
depth, and fluid overpressure with depth constrain the
dynamical model. We present model results for the Navarin
Basin COST No. 1 well including reconstructions of burial
history, fluid flow history, thermal history and
hydrocarbon generation history.
We also demonstrate how important geophysical
variables (such as permeability, porosity, fluid pressure,
fluid-flow rate, and thermal maturity indicators) vary
with depth and time. Comparison of the model results with
observed data illustrates and emphasizes the capabilities
of the modeling procedure.
The significance of the dynamical model is that it
permits a quantitative assessment to be made of (i) the
timing and depth locations of the generation and migration
of hydrocarbon in a basin, relative to the formation of
structural and stratigraphic traps, (ii) the timing of
the production of overpressuring and fracturing within a
basin, and (iii) the effect of cementation and dissolution
on the retention of hydrocarbons in a trap. The model
also enables an assessment to be made of the most likely
prospective areas for hydrocarbon accumulation in a
basin.

I.   Introduction. Thermal degradation through
geologic time of organic material in fine-grained

*Dept. of Geological Sciences, University of South Carolina, Columbia, SC 29208.

sediments is generally accepted as the dominant hydrocarbon generation mechanism (Hunt, 1979; Waples, 1984). That single statement triggers a plethora of problems that must be addressed if we are to improve our chances of finding hydrocarbon-bearing reservoirs. The problems can be split conveniently into three major groups: burial history, thermal history, and hydrocarbon generation, migration and accumulation history. As we shall see these groups are not isolated but are very much interdependent.

We examine each of the three groups, in turn, both in a general sense and for a particular well (Navarin Basin COST No. 1 well), in order to illustrate both the controls exerted by data and quantitative modeling capabilities in improving geologic understanding of the evolution of sedimentary basins in an objective fashion.

II. Burial History. A. General. Put to one side for the moment the question of hydrocarbon production and concentrate on dynamical evolution. At deposition sediments typically have porosities in the 50-75% range. When drilled at depth, say 15,000 ft., porosities are typically of order 10% or less. Thus an enormous amount of water has escaped from the sediments during their deposition and later evolution (For the example above, a 15,000 ft. section all at 10% porosity would have been 27,000 ft. thick if rehydrated to 50% porosity, so 12,000 columnar feet of water have been "lost"). Because of the fluid escape, the grain-to-grain contact pressure must increase in order to support the overlying sediment weight. Clearly a problem of fundamental concern, underlying all other aspects of basin analysis, is to understand the evolution of compaction in a sedimentary basin, and to track the dynamical evolution of fluid escape. In addition, the sediments must be supported on an underlying basement rock. As sediment is added to a basin, the basement will also respond to the applied load (rough values are 0.5 psi/ft. for water loading 1, psi/ft. for fully-compacted rock loading). If the basement material behaves as an incompressible fluid, then full Airy (1855) isostatic compensation occurs creating a surficial volume accommodation for further sediments. However if the basement behaves as a rigid unmoving plate, then no response occurs. Reality must lie between these two extremes, leading to basinal evolution models which encompass tectonic effects due to flexural plate behavior, tectonic rifting, wrenching, compression, and so on. Dynamical fluid escape depends intricately upon the permeability behavior of the deposited, and evolving, lithologic units. (Rough magnitudes of permeability are about a darcy for sand and millidarcies to microdarcies for shales (1 darcy $\approx 10^{-8}$ cm^2). As fluid loss proceeds porosity decreases, so permeabilities become smaller leading to an ever increasing delay in extracting the residual fluids. The addition of more sedimentary overburden is then compensated for by an increase in the retained fluid pressure over the hydrostatic value,

leading to an overpressured situation. Rough estimates
for fluid escape rates are about 50-200 ft./Myr from a
shale of millidarcy permeability at 0.2 psi/ft.
overpressure (i.e. 0.2 psi/ft in excess of hydrostatic
pressure gradient). If a sedimentary deposition rate of
greater than, roughly, 100 ft./Myr occurs then there is a
very strong possibility of significant overpressuring with
geologic time - again dependent on lithology. Since, on
average, about 75% of all deposition is shale,
overpressure is not a minor, rarely occurring phenomenon,
but rather has a major, dominant role to play and must be
allowed for in burial history calculations. If
overpressure occurs at any time then several consequences
are immediate. First, the compaction of overpressured
formations is less than without the overpressure, so the
formations retain thicknesses closer to their depositional
value for a longer period of time. In turn this effect
changes the structural and stratigraphic shaping of
sedimentary units, as well as providing a wider
(laterally) thoroughfare for fluid migration. Second, the
higher porosity relative to that of a non-overpressured
region lowers the thermal conductivity of an overpressured
region. In turn this increases the vertical temperature
gradient (often by as much as about 1C/100 m from the
"normal" value of around 2-2.5C/100m) in order to preserve
the heat flux. Thus overpressured regions tend to be
hotter than non-overpressured regions at the same sub-
surface depth (Hunt, 1979). Third, as a consequence of
the increased temperatures, the solubilities of carbonates
and sulfates (dominantly calcium, magnesium, barium) are
lowered, leading to calcite, gypsum, marl and anhydrite
precipitation which tend to strengthen formations by
diagenetic cementation. Fourth, isolated sand bodies in a
surrounding shale matrix will also be overpressured so
that a parametric geometry dependence on overpressuring
exists. Fifth, the thermal expansion of hot water in the
overpressured formations also provides an excess pressure
acting to divert fluid flow, and to pressure the
encompassing rocks. Since rocks have a finite mechanical
strength, if the total fluid pressure exceeds a limiting
value the rock must fracture thereby releasing fluid.
Empirically the total fluid pressure limit appears to be
roughly around 0.8-0.85 psi/ft. (excess fluid pressure of
0.3-0.35 psi/ft.) (Lerche, 1990). Thus the dynamical
evolution of sediments is firmly tied to the dynamical
fluid flow. As deposition continues fluid pressure
builds, rocks fracture and then reseal, pressure rebuilds,
leading to on-going episodic pulses of overpressure with
time. The ability to account for this behavior is a
crucial ingredient of any model.

    In addition, an increased temperature will also speed
hydrocarbon generation from kerogen, providing both
increased conversion from solid material to liquid (oil)
and gaseous components. These non-solid components
provide a generated pressure, can flow either in solution
with the prevailing waters or as separate phases, and so
impact on the dynamical burial history of the sediments.

A further, occasionally dominant, interaction occurs from clay diagenesis as smectites dewater to illites and kaolinites releasing their "bound" water into the fluid system (Hunt, 1979). The illitization is temperature and pressure dependent and is triggered by potassium catalysis but can be catalytically "poisoned" by sodium and calcium ions. Hence the dissolved minerals and ions in the subsurface waters also have a role to play in influencing sedimentary evolution.

Apart from fluid effects, major components in burial history evolution are also provided through depositional variation laterally in a basin, by erosional events, overthrusting, rotation of fault blocks, by intrinsic faulting of the sediments as syndepositional growth faults, tectonically induced faulting, or shear failure caused by differential stress on consolidated, lithified formations as the basin evolves. The presence of faults can lead to fluid escape up the faults, if the faults stay open for any length of time, thereby influencing the sediment evolution.

In addition to sand and/or shale two other major lithological types of rock fabric, salt and carbonates, also can be dominant controls on basinal evolution. Salt normally occurs in an evaporitic basin which is later open to clastic input. The evaporation can leave behind great thicknesses of salt (in the Gulf Coast of the USA estimates of the original Louann (Jurassic) salt thickness are in the several thousand feet range), overcoated with evaporites. Later sediment deposition on the underlying mobile, constant density, salt causes salt movement laterally and vertically giving rise to sediment distortion, faulting, and changes in the pattern of later sediment deposition through topographic expression from the buoyant salt and salt dissolution (Lerche and O'Brien, 1987). Indeed in some places (e.g. the Red Sea) evaporation is so rapid that salt is precipitating today on top of sediments.

Carbonates play a different role (Milliman, 1974). Deposited as the skeletal remains of marine organisms, carbonates basically "grow" by secretion of calcium. Hence carbonate reefs, platforms, pinnacle reefs, etc., represent situations of shallow marine deposition in warm water where nutrient supply is abundant and the optical depth to sunlight penetration is high. Large, laterally extensive, carbonate deposition is possible because the weight of the deposited carbonates causes basinal sag leaving a volume accommodation for further carbonate deposition. Indeed in the Dampier Basin of Western Australia upwards of 3,000 ft. of carbonate material, several tens of miles wide and about 100 miles in length was deposited in the Tertiary. Apart from the continuing impact such a weight has on the evolution of basinal structure, the stress on neighboring formations can often be sufficient to produce major faulting. In addition, the porous carbonates can act as good reservoir sites, can alter further depositional patterns of sedimentation, and can act as aquifers for ground water transport into the deep sub-surface, and for basinal deep waters to reach the surface.

In any attempt to determine the dynamical burial of sediments in a basin we have four major controls at each well, or pseudo-well, location. These controls are (a) the observed formation thicknesses, (b) the observed porosity with depth, (c) the observed fluid pressure with depth, and (d) the observed formation permeability with depth. To be acceptable a model must obtain agreement and consistency with at least these four controls at each location. It is not always easy to constrain models to honor these control criteria.

B. Description of the Model. A detailed mathematical description of the model can be found in the appendix to this paper. Here only a brief precis of the model is given.

The fluid flow/compaction model consists of three parts: a geohistory model, a thermal history model, and a hydrocarbon generation model. Because the model simulations are one dimensional, the input data for the model are those commonly used: geological and geochemical data from a single well, which makes the simulation useful in frontier areas where only a few wells are commonly available.

Fig. 1 shows the flow diagram of the one dimensional compaction model. The geohistory model reconstructs the burial history, basement subsidence, vertical fluid flow and the changes of porosity, permeability, pressure and fluid flow rate with both time and depth. Also the evolution of cementation, dissolution and fracturing caused by abnormally high pore pressure are simulated in terms of the change in formation permeability. The data required to run the geohistory model are the depth and age of each formation base, the solid mass (derived by inputting formation thickness, porosity and then computing the solid mass), and lithology of each formation, and the paleowater depth (Cao et al., 1986).

In the geohistory model, the simulation of sediment compaction is crucial because compaction of sediments either dominates or influences all other aspects of basin evolution. While the effects of diagenesis on compaction, porosity and permeability are undoubtedly significant, they are subordinate to the effect of overburden weight causing mechanical compaction. A primitive method of including cementation and dissolution is given in the Appendix, but is not developed further in this paper. Many processes such as burial history, thermal history, hydrocarbon generation (Fig. 2), migration and accumulation, etc. are either directly or indirectly dependent upon the sedimentary compaction.

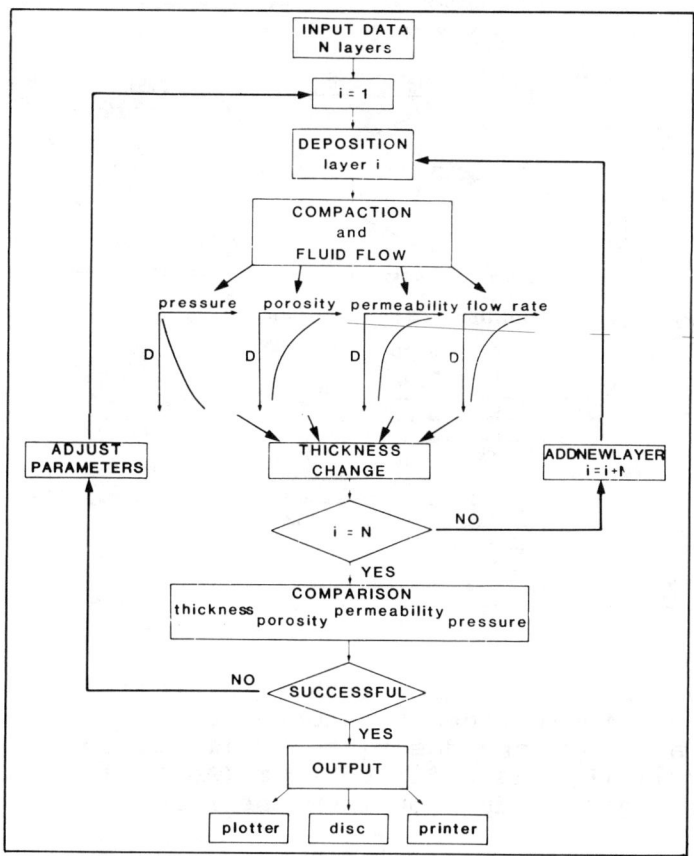

Figure 1.   Flow diagram for numerical implementation of the one-dimensional fluid
flow/compaction model.   The step labeled
"ADJUST PARAMETERS" represents the sequential
adjustment of the exponent indices A and B for
frame pressure and shale permeability as
explained in the Appendix.

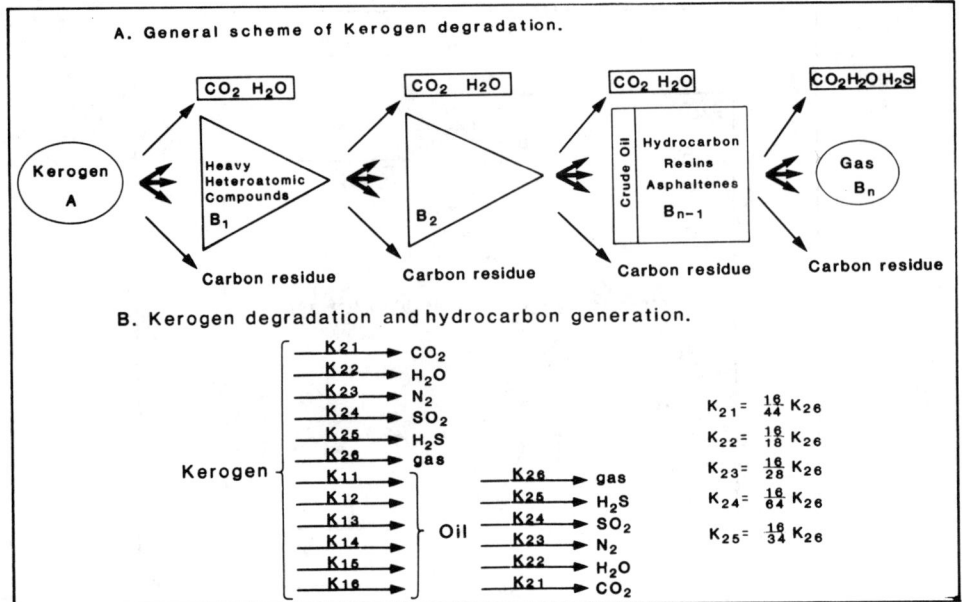

Figure 2. Schematic of chemical reactions used
to describe kerogen breakdown to oil and gas.
(A) as shown and described by Tissot and Welte
(1978); (B) as modified from (A) to incorporate
direct gas production from the kerogen complex.

        C. Navarin Basin COST No. 1 Well. For the Navarin
COST well, most of the detailed geological, geophysical
and geochemical data are reported in Geological and
Operational Summary, Navarin Basin COST No. 1 Well, Bering
Sea, Alaska, by Turner at al., 1984 (OCS Report MMS 84-
0081). Navarin basin COST No. 1 well, completed on
October 22, 1983, was the first well drilled in the remote
Navarin basin, Bering Sea, Alaska (Fig. 3). The Navarin
basin, a Tertiary sedimentary basin on the outer
continental shelf of the Bering Sea, is nearly 483
kilometers (300 miles) west of mainland Alaska and covers
an area about the size of Maine 83,400 sq. kilometers
(32,215 sq. miles). The northwest boundary of the basin
lies within the disputed area along the international

border between the U.S.A. and the Soviet Union. Shelf
water depths range from less than 97 meters (320 feet) to
152 meters (500 feet) at the shelf break. On the southwest
the basin is delimited by the northwest boundary
continental shelf break, and by shallow Mesozoic basement
rock on the southeast, northeast, and northwest (Steffy et
al., 1985).

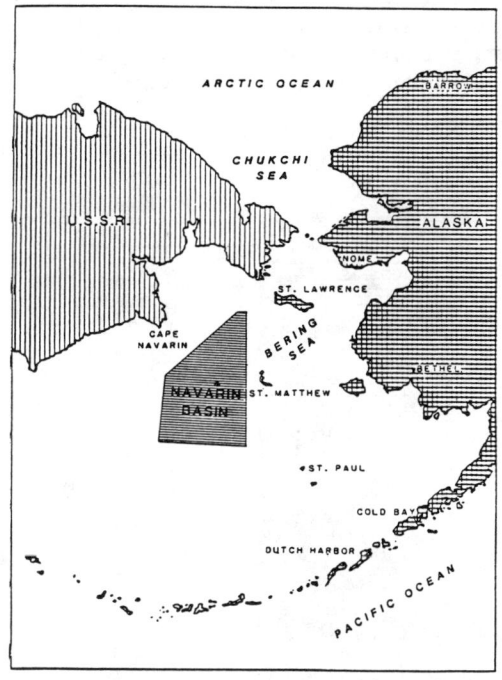

Figure 3. Location of the Navarin Basin COST
No. 1 well.

     The Navarin basin COST No. 1 well was located in the
east central part of the basin, reaching a depth of 4,999
m. The sedimentary section penetrated by the COST well is
composed of Neogene and Paleogene clastic sedimentary
rocks deposited in a marine environment, and Cretaceous
clastic sediments deposited in fluvial, flood plain and
deltaic environments. The Cretaceous and Tertiary
sedimentary rock are separated by a diachronous angular
unconformity. In the interval between 3,901 m and 4,587
m, late Cretaceous age rocks are intruded by numerous
basaltic and diabasic sills. Intrusion probably occurred
during early Miocene and late Oligocene time (Turner et
al., 1984).
     The penetrated section is divided into nine zones,
designated A through I on the basis of lithology,
deposition environments, diagenetic alteration, and well
log characteristics (Fig. 4). Lithologic zone A consists
of poorly sorted, silty, sandy mudstone and diatomaceous
ooze. Zone B consists of bioturbated, muddy, very fine
and fine-grained sandstone interbedded with sandy
mudstone. Zone C is composed of fine-grained, and very
fine-grained, muddy sandstone and siltstone interbedded

with mudstone and claystone. Zone D is characterized by
sandy mudstone, fine-grained muddy sandstone, and
claystone with rare lenses of siltstone and sandy
carbonate. Zone E consists of poorly sorted gray
claystone, mudstone, and sandy mudstone with abundant
detrital clay matrix. Zone F consists of dark-gray
calcareous claystone and sandy mudstone. Zones G-H are
characterized by siltstone, very fine grained sandstone,
mudstone, claystone and coal, which is intruded by diabase
and basalt sills. Zone I is composed of claystone,
siltstone, tuff and mudstone deposited in a marine
environment.

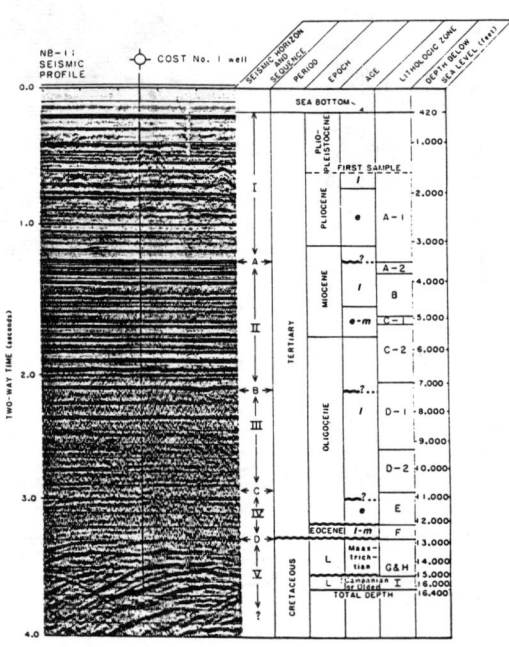

Figure 4. Seismic profile across the region
containing the Navarin Basin COST No. 1 well
and corresponding geologic column from the
well.

There is a major diachronous angular unconformity
between late Cretaceous and younger Tertiary sedimentary
rocks. The vitrinite reflectance data are abnormally high
in the late Cretaceous section below the unconformity
because of the thermal intrusion, which makes it difficult
to simulate thermal history of these sediments by the
inverse method presented in this paper. Accordingly the
inverse model is applied only to the sequence above the
unconformity, i.e. the Cenozoic rocks.
Table 1 lists the basic input data of the Navarin
COST No. 1 well. The Cenozoic sequence is split into 13
modeling layers from bottom to top, based on lithology,
paleontology, biostratigraphy and deposition environments.
Layers 2 and 4 represent unconformities and are treated as
hiatae, marking eustatic events (Turner et al., 1984).
The temperature at the sediment-water interface and the

bottom hole temperature are based on the temperature data of Fig. 5.   The vitrinite reflectance data with depth are shown in Fig. 6.   The kerogen content data are generalized from the geochemical data in Turner et al. (1984).

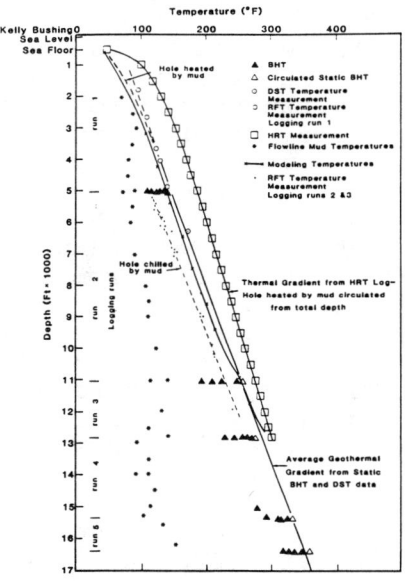

Figure 5.   Several measured present-day temperature profiles with depth are presented as explained in the figure legend. In order to obtain a present-day heat flux we used the modelling temperature with depth curve shown.   This curve was obtained to satisfy the best average geothermal gradient.

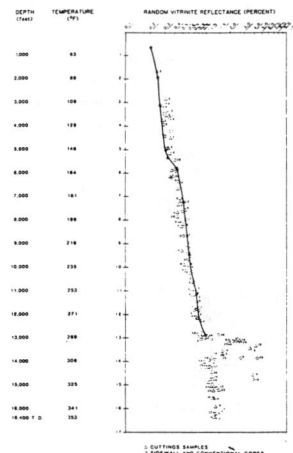

Figure 6.   Plot of vitrinite reflectance observations with depth (symbols). Super-imposed is the solid curve representing the predicted vitrinite reflectance obtained by the thermal inversion pro-cedure explained in the Appendix.

D. Comparison of the model results with real data.
It took 80 seconds CPU time to run the COST No. 1 well on
the mainframe IBM 3081 at the University of South
Carolina. Based on the data available, the following
parameters have been used to make comparisons of the model
results with the observed data.

1) Thickness. Table 2 gives the formation thickness
changes of each modeled layer during compaction. The
table shows that the thickness of each layer decreases
after its deposition time, and that the compaction rate is
different at different compaction stages. At the early
(shallow depth) stages of compaction, the sediment
thickness decreases faster than at the later (deeper
depth) stages of compaction. For example the compaction
rate of layer 1 (shale) decreases from 4.25 m/Myrs at the
end of layer 4 deposition to 2.71 m/Myrs at the end of
layer 10 deposition. The total modeled thickness of the
whole sequence at the present day is 3,729.0 m, and the
true thickness is 3,737.7 m, with a relative error of
about ±1% for individual layers. Table 2 shows similar
accurate predictions of layer thicknesses for all
subsequently deposited layers.

2) Porosity. Fig. 7 shows present day porosity
versus depth for the Navarin COST No. 1 well. The solid
line is the modeled porosity curve and the points are
measured porosity values for 114 samples from sidewall and
conventional cores. No measured porosity data are
available in the interval less than 914 m deep. It is
apparent that the modeled porosity fits the measurements
rather well from 914 m to 3,962 m. Table 3 gives a
comparison of modeled porosity and measured porosity for
each layer, showing a maximum fractional error of 17% and
an average fractional error of order 10%.

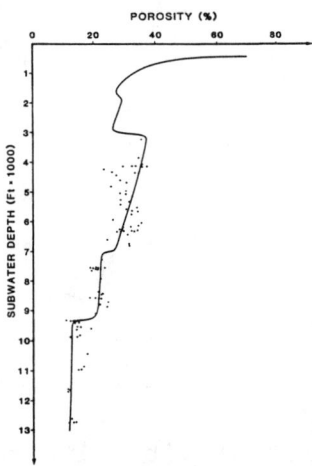

Figure 7. Porosity measurements with depth
in COST No. 1 (shown by dots) and the predicted
porosity variation (solid line) obtained by
running the fluid flow/compaction code.

3)   Formation Pressure.   One of major differences
between an isostatic model and the dynamic model used here
is that the dynamic model can simulate the evolution of
formation pressure.   It is important to examine the degree
to which the modeled formation pressure with depth is in
accord with the measured formation pressure.   The
existence of abnormally high pore pressure in the Navarin
COST No. 1 well has been shown by wireline logs, formation
drill stem tests and the drilling exponent method (Turner
et al., 1984). The wireline log and formation test data
clearly identify a major zone of abnormal formation pore
pressure in the interval below 2,874 m in the Navarin
Basin COST No. 1 well. A shallow zone of excess pore
pressure between 762 m and 1,170 m also appears to be
present.

Figure 8 shows the pressure profile for the Navarin
Basin COST No. 1 well. The thick dashed line is the
modeled formation pressure curve showing two abnormal pore
pressure zones. The deep zone starts at a depth of about
2,804 m and extends to the intrusive unconformity. An RFT
test at 3,098 m (subsea depth) within the zone yielded 362
atm (5,320 psi) on pressure build up (i.e. 49 atm (726
psi) greater than the anticipated 313 atm (4,594 psi) if
the well had been normally pressure). Our modeling
pressure at the same depth is 358 atm (5,256 psi), 45 atm
(662 psi) greater than anticipated for a normally
pressured section. The shallow abnormal pressure zone is
predicted by our model to occur in the interval of 305
meters (1,000 feet) to 1,158 meters (3,800 feet) - as seen
in Figure 8. (Details on the effective stress law used are
provided in the Appendix).

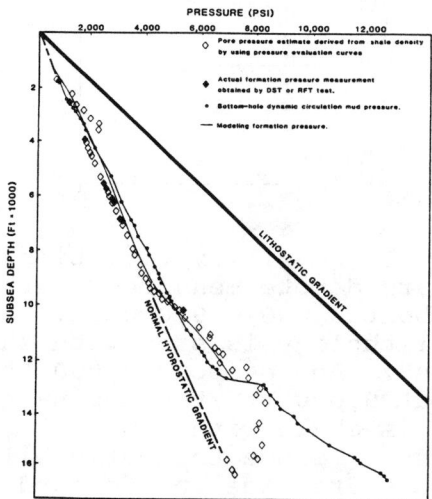

Figure 8.   Pore pressure measurements with
depth. The three different symbols represent
different methods of estimating pore pressure
as explained in Turner et al. (1984).   The
light solid curve is the predicted variation
of pore pressure with depth obtained from
running the fluid flow/compaction code.

4) Permeability Variations with Depth. Figure 9
shows the logarithm (in md) of measured formational
permeability variations with depth (asterisks) and
superimposed on the figure is the predicted permeability
variation using the fluid-flow/compaction model which
allows for dynamical evolution (see the Appendix for
details of the permeability law being used). To be noted
from figure 9 is the comparison of the predicted variation
and the measured values - in a general trend sense. In
particular the agreement between 914 meters (3,000 feet)
and 3,048 meters (10,000 feet) is more than at greater
depth. Below about 3,048 meters (10,000 feet) the
dominant lithology is shale with thin interbedded sand
layers. The measurements of permeability below 3,048
meters (10,000 feet) are, of course, done on the sand
layers, as potential hydrocarbon reservoir sites. From
the observed overpressure we are fairly certain that the
bulk permeability is quite small below 3,048 meters
(10,000 feet) since overpressuring continues to increase.
Hence the sand permeability measurements can be mis-
representations of the over-all variation of bulk
permeability, which is controlled by the dominant shale
lithology.

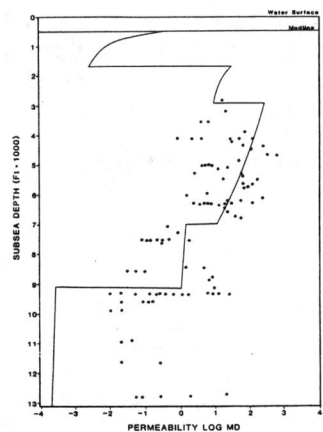

Figure 9. Permeability variations with
depth. Stars denote measured values.
Between about 2,000 - 9,000 ft, the
dominant lithology is sand with shale
"stringers". Above about 2,000 ft and
below about 9,000 ft, the dominant
lithology is shale with sand
"stringers".   Since the permeability
measurements are made in the sand
"stringers" below about 9,000 ft,
they misrepresent seriously by a
large factor the bulk permeability
of the dominant shale unit. The
solid line is the predicted variation
of permeability with depth obtained
from running the fluid flow/compaction
code.

Above 914 m there are, apparently, no permeability measurements, but the shallow zone of overpressuring observed, and also predicted by the fluid-flow/compaction model, indicates that the permeability must be fairly low in order to maintain the shallow overpressure. The predicted variation of permeability in the shallow $\leq$ 610 m zone is, therefore, consistent with the overpressure observations.

When it is borne in mind that the individual formation thicknesses with depth, the porosity variations with depth, the fluid overpressuring with depth, and the permeability variations with depth are all correctly predicted (with only two "free" numbers adjusted with depth to minimize discordances between observations and predictions, see the Appendix for details of our procedure), it would seem that the fluid-flow/compaction modeling provides a ruggedly stable tool capable of being used in a predictive capacity with a relative precise degree of accuracy in predicted bulk properties.

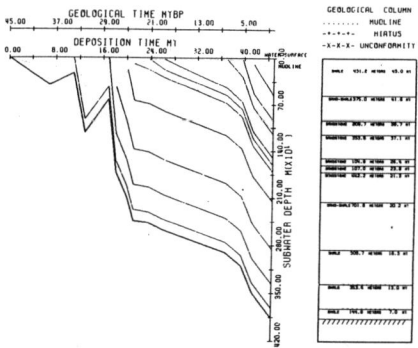

Figure 10.   Reconstructed geohistory of COST No. 1 well allowing for overpressuring with time. Note the rapid subsidence event at about 28 Ma, with relatively slow subsidence thereafter, and two major unconformities prior to the rapid subsidence.

Fig. 10 shows the geohistory of the Cenozoic sequence in the Navarin Basin COST No. 1 well. Since the deposition of dark-gray calcareous claystone and sandy mudstone in the Eocene, the basin has undergone two major uplifts: one occurred between the Eocene and the early Oligocene, and the other occurred at the early Oligocene (Turner et al., 1984). During these two uplifts coarse-grained sediments were deposited in shallow water. After the early Oligocene uplift, the basin subsided quickly during the rest of the Oligocene, and was followed by a slow subsidence with a nearly constant subsidence rate during the Miocene, when the depocenter was depositing

dominantly regressive sandstone, which forms the most
promising reservoir rock (Steffy et al., 1985). From
Pliocene to present, the basin has undergone subsidence
with fine-grained material deposition. Fig. 11 shows the
basement subsidence with time, assuming a basement depth
of 9144 m (Steffy et al., 1985). The dark line in Figure
11 is the total basement subsidence, the dotted line is
the basement subsidence caused by tectonism. It can be
seen that the tectonic basement subsidence is dominant
from Eocene to Oligocene, and the sediment loading effect
played a more dominant role thereafter in influencing the
basement subsidence. Fig. 12 gives the rates of total
basement subsidence, tectonic basement subsidence and
sedimentary deposition with time. The curves of Fig. 12
show a maximum rate during Oligocene and a nearly constant
rate during the Miocene.

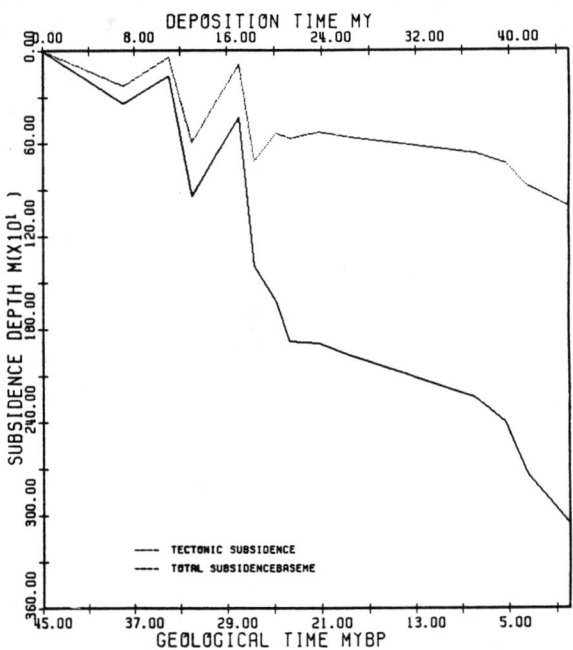

Figure 11. Curves of basement subsidence
with time. The solid curve represents the
total basement subsidence including the
effects of sediment loading, the dotted
curve represents the tectonic subsidence
when the isostatic compensated subsidence
due to sediment load is removed in the
manner of Steckler and Watts (1978).

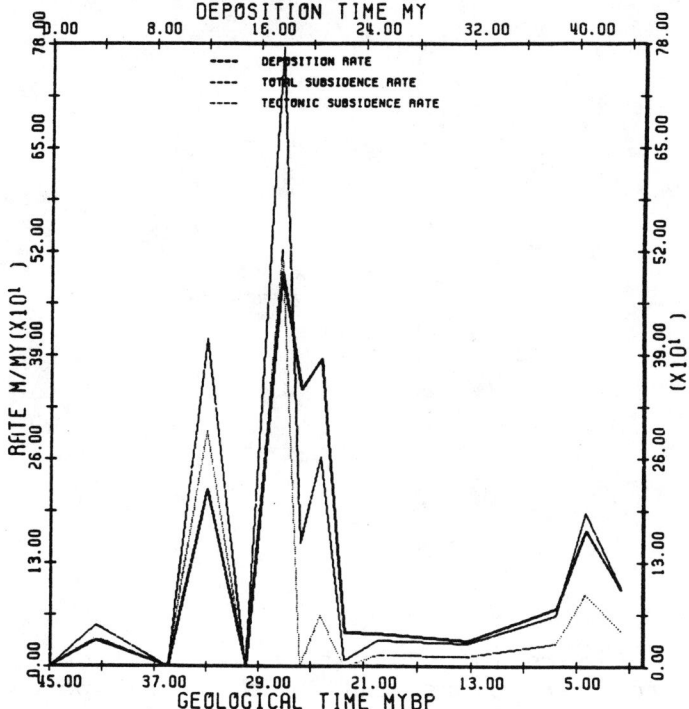

Figure 12.   The three curves show (a) the
rate of subsidence including sediment load
(light dashed line):   (b) the rate of
basement subsidence with the isostatic
compensation due to sediment load removed
(heavy dashed line), and (c) the sedimentary
deposition rate (light dotted line).

        In summary: examining the whole history of compaction
in the COST No. 1 well, we find that the Eocene shale
(zone F) and early Oligocene shale (zone E) were
overpressured soon after their deposition and these zones
are predicted to have undergone fracture formation during
the late Oligocene. Indeed, Fig. 13 shows the profiles of
porosity, permeability, pressure and fluid flow rate with
depth at the end of layer 6 deposition (24.84 Myrs BP).
There is then a fracture zone at around 1,400 m with a
"jump"   in permeability and fluid flow rate. The "jump"
persists until the fluid pressure drops below a critical
value (set at about 0.85 of lithologic overburden) when
the fractures are assumed to shut.

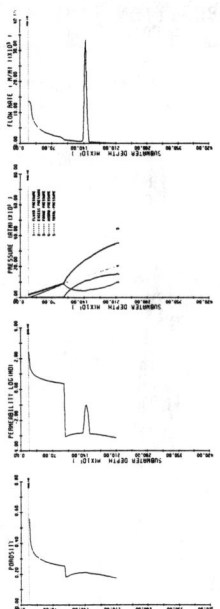

Figure 13. Illustrative profiles of predicted variations with depth of (a) porosity,
(b) permeability, (c) pressures, and
(d) fluid flow rate at 24.8 MYBP. Note
the pressure of a relative 'spike' in
permeability at about 1,400 m depth
representing the fracturing of sediments at that depth at that time. The
fluid flow also shows a 'spike' at the
same depth, corresponding to a total fluid
pressure in excess of 0.85 of the lithologic overburden.

III. Thermal History. A. General. Put to one side
for the moment the problems of dynamical burial history
and hydrocarbon production and concentrate on the thermal
history of the sediments. As we shall see in the next
sub-section, knowledge of the thermal history of the
sediments is a crucial ingredient in assessing the
hydrocarbon generation and migration history. On the
assumption that the dynamical burial history has been
determined precisely, we need to determine the temperature
history of each and every sedimentary unit with time and
burial depth. In turn, determination of the temperature
history rests on two fundamental factors: a determination
of the sources of heat flux through, and within, the
sediments, and how those heat fluxes change with time and
spatial location; and a determination of the transfer of
heat (conduction, convection) through the sediments.
Significant sources of heat internal to the sedimentary
sequence can be highly radioactive layers (e.g. the
Kimmeridge clay in the North Sea basin), igneous
intrusives, frictional heating by overthursting, and so
on. By and large it would appear that such internal heat

sources, while often dominant locally, are secondary in magnitude and regional character to heat flux variations through the basement rocks.

Present day estimates of sedimentary rock thermal conductivities vary in the rough range 7 ± 3 mcalcm^{-1} sec^{-1}C^{-1} with the low values being for high porosity unconsolidated rocks and the high values for low porosity, nearly fully compacted sediments (Hunt, 1979). Ignoring lithology effects, a rough quasi-empirical estimates connecting fractional porosity, ϕ, and bulk thermal conductivity can be obtained from (Lewis and Rose, 1970)

$$\kappa = K^{1-\phi} K_w^{\phi}$$

where K_w is thermal conductivity of water ($\sim$1 mcal cm^{-1}sec^{-1}C^{-1}) and K_R is the thermal conductivity of fully consolidated sediments ($\sim$10 mcal cm^{-1}sec^{-1}C^{-1}). Present day temperature gradients are around 2C/100 m so that, if conduction dominates, estimates of basement heat flux today are around $Q = 1$ HFU[1] (1 HFU = 42 Wm^{-2}). A dynamical range of about 0.5 - 3 HFU is not inappropriate for most sedimentary basins today. Thus while we can estimate from measurements the spatial variation of heat flux in a basin today, we have to determine how it varied in the past. Two end-member approaches are possible: (i) we can be guided by our knowledge of the dynamical evolution of a basin, posit a model of basement motion and a model relation between basement motion and basement heat flux variations, determine any parameters in the proposed basement motion, and apply these directly to the heat flux. Such schemes give rise to a variety of popular models of the stretching-rift sort (McKenzie, 1978), to phase change models (Falvey and Middleton, 1981) and the like. Most such models end up predicting some sort of high heating in the past ($Q \approx 4$-10 HFU) with a cool-down to the present day. The weakness here is that both a dynamical/thermal model connection must be made, as well as a dynamical model from which to extract salient parameters; (ii) we can be guided by the fact that various thermal indicators exist in wells today. These thermal indicators usually vary in some systematic way with depth. The general sense of argument here is that if the temperature dependent physics/chemistry of the indicators along the known burial paths is given then the evolution of temperature with time can be determined. For instance Lopatin (1971), and later Waples (1980), have popularized the idea that vitrinite reflectance, $R(z)$, measured with depth z, can be related to a time-temperature integral (TTI) of the form

$$TTI = \int_o^t \exp\left[T(t') - T_c)/T_D\right] dt'$$

where the integral is taken along the burial path of the sediments from deposition to the present day, and T_c and

---

[1] HFU = Heat Flux Unit

T_D are two scaling constants. The basic idea is to make TTI a hydrocarbon measure. To foster such ends it has been proposed (Waples, 1984) that T_C is best set at 105°C, and T_D at $10/\ln_e 2$°C. But a connection still needs to be found between R(z) and TTI. If the geothermal gradient is constant for all time, then TTI can, of course, be calculated for each sedimentary layer (but not measured) and plotted versus the measured R(z) values. A quasi-empirical relation is then generated connecting TTI and R. Unfortunately there is no way of knowing whether the geothermal gradient is constant with time, nor is there any way of assessing the choice of parameter values for T_C and T_D. The point is that there is no fundamental requirement for the maturity of vitrinite reflectance to be slaved one-to-one to that of hydrocarbons.

What is needed is a physical/chemical equation connecting vitrinite reflectance to its own TTI (and not that TTI purported to represent hydrocarbon maturity). Then we stand a chance of being able to determine both the equivalents of T_C and T_D, as well as the heat flux variation in time. This same point has also recently been made by Waples (1988) in a strong plea for kinetic descriptions to replace the Lopatin (1971) TTI. Every thermal indicator (vitrinite, steranes, pollen translucency, hopanes, Ar^{39}/Ar^{40}, apatite fission tracks.....) has its own particular TTI, each related in divers ways to the especial nature of the individual indicator. As a consequence inverse schemes, based on multiple thermal indicators, are rapidly replacing the primitive thermal maturity procedure of "calibrating" the Lopatin-Waples TTI to vitrinite reflectance (for which values of TTI = 15 and TTI = 75 are often taken as marking the onset of entry to the "oil window" and "gas window" respectively (or equivalently R = 0.6 and R = 1.1)).

Suppose that such devices have provided a basement heat flux variation with time at each location in a basin. We then have to be concerned with a good determination of paleo-temperature evolution at each location. Two major problems still confound the issue.

First is the problem of assessing the thermal conductivity variation in the basin because lithology and porosity differences spatially will act as focusing and defocusing domains for heat transport by conduction. Second is the problem that high permeability sedimentary beds can provide excellent pathways for convective transport of heat carried by flowing fluids, thereby preferentially heating and/or cooling regions in amounts considerably different than conduction alone could provide. Hydrocarbon generation is affected, as is hydrocarbon transport and further generation of oil to gas after location in a reservoir site, by such convective heat transfer. In addition there is a change in mineral solubility by high temperature fluid flow leading to cementation and/or dissolution which, in turn, influence the migration pathways by diagenetic evolution. In turn the transport of heat impacts on fluid expansion and rock failure by pressure, as well as on the porosity. But the

porosity variations relate directly to the thermal
conductivity, thereby altering the temperature gradient
which, in turn, changes the convective heat transfer, and
also reflect directly on the burial history. Thus the
circle is complete - thermal and burial histories are
intimately intertwined and cannot be considered in
isolation but must be addressed simultaneously.

The only direct constraints that we have at the
present day on the thermal history are measurements of
multiple thermal indicators with depth as well as present
day temperature gradient. In order to be acceptable any
thermal evolution model must, at the least, be consistent
with all such data.

B.   Navarin Basin COST No. 1 Well Results. 1)
Temperature. Figure 5 shows the temperature profile for
the well. The modeled temperature variation with depth
fits the measured temperature data well. The modeled
temperature curve shows two high thermal gradient zones,
consistent with the abnormally high pressured zones
(shales) and the one normal pressure zone (sandstone).
Within the shallow over-pressure zone (above 1,158 m) the
average predicted thermal gradient is 4.0°C/100 m, (cf.
4.6°C/100 m in Turner et al., 1987); within the normal
pressure zone (from 1,158 to 2,804 m) the average
predicted thermal gradient is 3.3°C/100 m, and within the
deep overpressure zone the average predicted thermal
gradient is 4.7°C/100 m. In Turner et al. (1984) an
average thermal gradient of 3.2°C/100 m from 1,158 m to
4,999 m was estimated without considering the thermal
gradient change in the deep overpressured zone. The
reason for the higher thermal gradient in the
overpressured zones is that the thermal conductivity of
overpressured sedimentary rocks is lower than that of
normal pressured sedimentary rocks because the
overpressured zone has an higher fluid content and water
has a low average thermal conductivity than fully
compacted matrix. Hence the temperature gradient
increases in order to balance the heatflux through the
sediments at each instant of time.

2) Vitrinite Reflectance. Fig. 6 shows the
vitrinite reflectance profile of the well. The dark line
is the predicted vitrinite reflectance with depth stopping
at the unconformity based on the inverse methods mentioned
earlier and detailed elsewhere (Lerche et al., 1984)
providing a variable heat flux with time. The curve
generally reflects the change of measured vitrinite
reflectance with depth above the unconformity. To predict
the abnormally high vitrinite reflectance below the
unconformity an intrusion thermal model is needed -
something we are currently giving a large amount of effort
to in our attempts to quantify using inverse techniques.

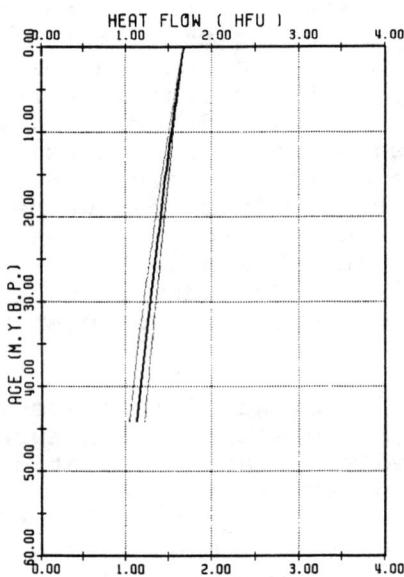

Figure 14. Variation of heat flux with
time obtained by thermal inversion of the
vitrinite reflectance values recorded in
the well, along the burial paths of the
sediments. The heat flux variation obtained
is the best one parameter variation capable
of providing minimum discordance with the
observations, and represents a slow warming
trend in the region by about 0.7 HFU over
the last 45 Ma.

    3) Thermal history. Fig. 14 shows the heat flux
change with time, reflecting a warming-up trend with a
present heat flux of 1.7 HFU. Based on the modeled heat
flux change with time obtained from the burial history in
agreement with downhole formation thicknesses, porosity,
permeability, fluid pressure and vitrinite reflectance
measurements, we can then reconstruct the thermal
temperature history (Fig. 15), which shows a variable
thermal gradient both in the past and with depth. Based
on the temperature history in Fig. 15, in Figs. 16 and 17
we show the thermal maturity evolution of the Cenozoic
sequence using iso-reflectance lines as a maturity
indicator (figure 16) and Waples' TTI (Waples, 1980)
(Figure 17). The iso-vitrinite reflectance lines, plotted
in Fig. 16, show the thermal maturity of each layer. If,
for illustrative purposes, a vitrinite reflectance value
of 0.6 is chosen as defining the onset of the "oil window"
(Waples, 1980) then only the Eocene and early Oligocene
shales (Layers 1, 3, and 5) are thermally mature and
within the "oil window", younger sediments are immature.
However, Fig. 17 indicates that only layer 1 and layer 3
are mature and in the "oil window" if a TTI value of 10 is
chosen as the threshold (Waples, 1980). Also the depth
and time at which each layer entered the "oil window" can
be read from figures 16 and 17.

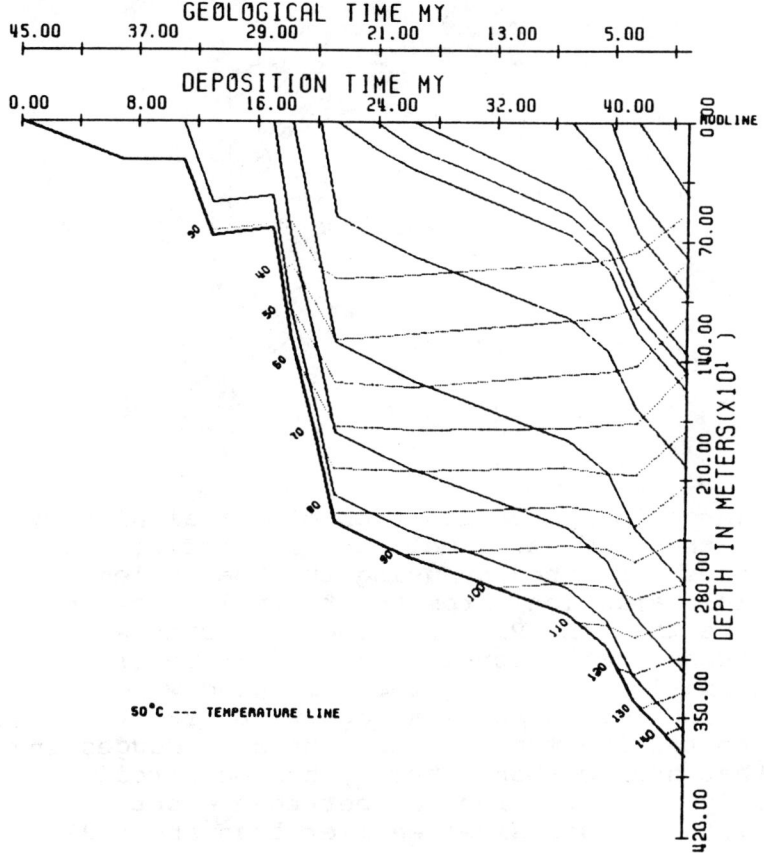

Figure 15. Superposition of burial history
of the COST No. 1 well and iso-temperature
lines, obtained by using the best paleoheat
flux variation from the theraml inversion.
Note that the 100C isotherm (occasionally
considered a maturation indicator) occurs
at about 2,800 m and starts at about 15 Ma,
indicating kerogen maturity at the present
day only for formations deeper than 2,800 m
which were deposited earlier than 29 MYBP.

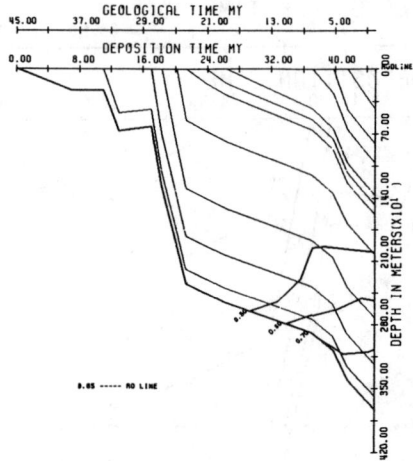

Figure 16.   Superposition of burial history
of the COST No. 1 well and iso-reflectance
lines, obtained by using the best paleoheat
flux variation from the thermal inversion.
Note that the R_o = 0.6 iso-reflectance line
(occasionally considered indicative of
formations entering the "oil window"),
varies from 2,800 m at about 20 MYBP to
about 2,500 m at the present day, suggesting
that hydrocarbon maturity has occurred
only for formations deeper than about
2,500 and deposited earlier than about 27
MYBP.

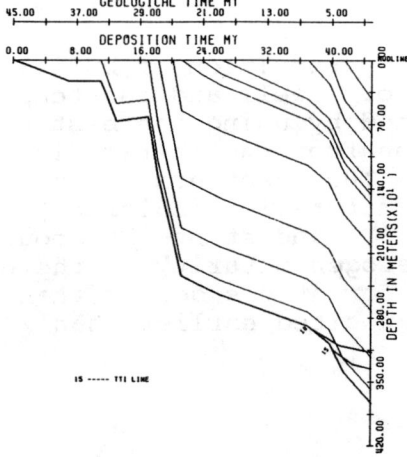

Figure 17.   As for figure 16 but with iso-TTI
contours. A TTI of 15 is often considered to
represent the start of oil generation and,
in this case, would suggest maturation only
for formations deeper than about 3,200 m
deposited earlier than about 30 MYBP.

IV.   Hydrocarbon Generation, Migration and
Accumulation Histories.   A.   General.   Put to one side the
dynamical and thermal histories and concentrate on the
hydrocarbons.   General considerations (Hunt, 1979; Brooks
and Welte, 1984) suggest that the dominant mode of
hydrocarbon production is by thermal degradation of
kerogen after sedimentary burial.   A "rich" source rock
might contain around 1-5% TOC (total organic carbon)
capable of producing hydrocarbons.   Three major problems
need to be addressed: (i)   how are hydrocarbons produced
from kerogen?   (ii)   how do hydrocarbons migrate out of
the (dominantly shale) source beds, and then how do they
move thereafter (primary and secondary migration)?   (iii)
what sort of accumulation sites and sealing conditions are
available for the hydrocarbons, and do those sites persist
through to the present day?

Hydrocarbon production is a peculiar phase change in
which solid material (kerogen) is turned into mobile
liquids (oils) and mobile gases (mainly methane but with
some ethane, H_2S, CO_2, N_2).   Furthermore, oils are virtually
incompressible and normally have a specific gravity around
0.8-1.01 - usually less than sub-surface brines (~1.0 to
1.3).   Gases are highly compressible, and normally have a
specific gravity of around 0.1-0.3 at production at depth.
Thus oils and gas are buoyant relative to water (buoyancy
pressure ~0.1 psi/ft. for oils, ~0.3-0.5 psi/ft. for gas).
Some oils (e.g. isopropyl alcohol) are completely miscible
with water, others are not (Hunt, 1979), while gas tends
to be completely miscible until the "bubble point"
pressure and temperature are reached when gas ex-solves
and transports as a separate phase at lower temperatures
and pressures.   It is these large phase change properties
that make the job of assessing potentially hydrocarbon
reservoir sites so difficult.

The ability to provide a detailed kinetic scheme
relating the thermocatalytic breakdown of kerogen to
hydrocarbons is fraught with major difficulties - perhaps
the worst of which is to provide a scheme handling organic
macromolecule thermocatalytic chemistry (kerogen) breaking
down to methane (molecular weight 16) in the sub-surface
when ionic, mineral, and surficial chemistry influence and
modify the chemical kinetics that otherwise might be
thought appropriate based on laboratory measurements.
From van Krevelen diagrams of H/C and O/C we recognize
(Hunt, 1979) that there are at least three types of
kerogen - presumably each having its own kinetic breakdown
scheme, and yielding different production efficiencies for
oil and gas generation.   We also know that as the sub-
surface temperature increases with depth oil gradually
converts to gas (Hunt, 1979).

Attempts to by-pass the detailed kinetics then
revolve around the idea of providing a kinetic breakdown
scheme which encompasses the gross scale of the
observations but not the details.   Thus questions
concerning the amount of oil generated are fair game, but
not questions concerning the amounts of n-butane
production versus paraffin wax, i.e., the kinetic

frameworks are "lumped impedance" types of devices
attempting to bracket the over-all trend of behavior.
Such a "lumped" kinematic scheme was introduced by Tissot
(1969), refined by Tissot and Espitalie (1975), and
popularized in Tissot and Welte (1978). Since then, many
other "lumped" kinetic frameworks have been suggested. A
common theme among them is that they all require multi-
channel reactions, in series or parallel, with
specification of many reaction rate constants and
activation energies which relate kerogen to oil and gas,
and oil to gas. All are temperature dependent and most
involve first order chemical kinetics.

By whatever scheme, then, various sets of
thermochemical kinetic breakdown equations are available
for assessing hydrocarbon production from kerogen.
Typical total production estimates range from 1-500 mg of
hydrocarbons per gm of kerogen, while typical estimates of
the onset of significant oil (gas) production rates are
around 100°C (150°C) respectively.

Having converted kerogen to mobile hydrocarbons the
question arises as to how the liquids escape the fine-
grained clastic sources beds of generally low permeability
(primary migration). Several mechanisms have been
proposed. Thermal expansion coefficients of fluid
hydrocarbons are larger than rock values. Hence the
conversion of kerogen to fluid puts an additional fluid
pressure on the encompassing source rock leading to the
possibility of microfracturing and fluid loss through the
high permeability fractures. An alternative suggestion is
that the volume occupied by produced hydrocarbons is
larger than that of the original kerogen leading directly
to an increase in pressure, again driving fluid motion and
microfracturing. A third suggestion is that capillary
pressure at sand/shale interfaces has a dominant role to
play. Typical pore radii in fine-grained materials are
around $10^{-5} - 10^{-6}$cm, while in sands the size is around
10^{-4} cm. The surface tension difference, T, between oil
and water is around 100 dyne/cm so that the capillary
pressure, p, at a sand/shale interface is $p \approx 2T$ $(1/r_{fine} -
1/r_{coarse}) \approx 20-200$ atmospheres ($\equiv 300-3000$ psi). Thus
capillary pressure is the largest force at the boundary
but drops rapidly away from the boundary. Hence oils tend
to displace water in coarse-grained beds abutting the
fine-grained source bed, while water displaces oil in the
source.

By whichever process, generated oil and gas now find
themselves in a relatively high permeability bed. They
can then move (secondary migration). Several forces act
on the hydrocarbons. Any excess fluid pressure in an
overpressured formation causes a hydrodynamic drive acting
not only on the ambient sub-surface fluids but also on the
hydrocarbons. That fraction of oil and gas which go into
solution move with the prevailing fluid. But the non-
soluble gas/oil fractions have a vertical buoyancy
pressure drive acting on them of around 0.1-0.5 psi/ft.
Resistance from overlying less permeable beds then means
that the non-soluble fractions tend to move along the path

of least resistance, which is not necessarily vertical. Thus the dynamical evolution of the structural aspects of bedding geometry now come into play; up-dip migration to structural culminations being favored by oil and gas, with gas motion attempting to outstrip oil motion (by about a factor of five to ten in speed) because of its greater buoyancy drive.

On the other hand it is not just the pressure gradient which controls fluid motion. Darcy's law notes that the speed of motion is proportional to the permeability as well as to the driving pressure gradient. As the hydrocarbons move from their source location to potential accumulation sites, any significant decrease in permeability will effectively bar their further migration. Such decreases can be caused by compaction as the basin develops, by diagenetic cementation, by spatial variations in lithology and, of course, by tectonic forces altering the basin. Thus as well as the desire of hydrocarbons to seek out structural highs as accumulation sites, there are also stratigraphic sites of potential accumulation which may, or may not, be associated with structural evolution and development. And the stratigraphic variations also develop with time.

Three further possibilities cause on-going variations in hydrocarbon accumulation sites: hydrodynamic versus hydrostatic conditions, sealing conditions, and continued generation of oil to gas in a reservoir.

Suppose that oil has succeeded in finding a place to accumulate at some time in a basin's history. Continued evolution of the basin can lead to aquifer flow at later times attempting to flush the oil from its original accumulation site. The hydrodynamic forces can flus the oil down-dip or up-dip depending on their direction. The prevailing milieu of physical dynamical transport plays a major role in some hydrocarbon regions, as has been amply demonstrated by Hubbert (1940) and Dahlberg (1982), both theoretically and with case histories.

In addition the sealing conditions around a filled reservoir at some instant of time, such as an overlying impermeable shale, can change significantly as later deposition and basinal development impact on the stress conditions of the seal - occasionally leading to fracturing and faulting of the seal and consequent loss of the hydrocarbons.

We also note that the models of kinetic degradation of kerogen to oil and gas contain components linking oil degradation to gas production. So an oil-filled reservoir, even if kept sealed, will through the course of time, gradually convert from oil to gas. But this conversion increases the pressure on the seal (because the gas buoyancy pressure, is at a minimum, about five times that of oil) leading to the distinct possibility of later rupture and hydrocarbon loss.

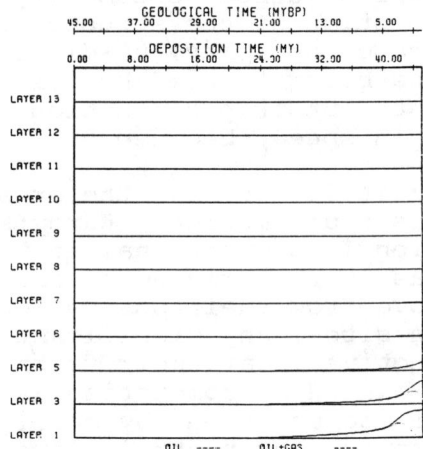

Figure 18. Hydrocarbon generation history
of various layers using the Tissot-type of
model supplemented by the fluid flow/com-
paction burial history and the variation
of heat flux with time obtained by thermal
inversion of the vitrinite reflectance data.
Note that only the deepest layers are
suggested to be in the oil window with
generation having started around 10-5 MYBP.

B.   Navarin Basin COST No. 1 Well Results.
Hydrocarbon Generation History. Fig. 18 shows the
hydrocarbon generation history of the Cenozoic sequence in
the Navarin Basin COST No. 1 well based on the temperature
history and burial history created in the model. The
figure indicates that the shales of Eocene and early
Oligocene are good source rocks in terms of thermal
maturity (layers 1 and 3), layer 5 has begun to generate
hydrocarbons and layer 6 and layers above are immature.
Figures 19 and 20 provide more details concerning
hydrocarbon generation for layers 1 and 3. The figures
shown are based on the hydrocarbon generation model
presented by Tissot and Welte (1978). Our modified model
results did not provide a significant discordance with the
Tissot-Welte results (and are therefore not shown). The
burial histories of layers 1 and 3 are shown in the left
hand side plots of Figs. 19 and 20, giving the changes in
the depth of the base of layers 1 and 3 with time. The
total amount of hydrocarbons generated for layers 1 and 3
with time are shown in the middle plots of Figs. 19 and
20, as is the transformation ratio of kerogen in layers 1
and 3. The right hand plots give the oil generation, gas
generation and kerogen transformation ratio with time.
From Fig. 19 we see that the Eocene shale (layer 1)
arrived at the "oil window" at early Pliocene (7 Myrs BP,
2,000 m), reached peak oil generation at late Pliocene
(3.0 Myrs BP, 3,500 m) with the peak gas generation
probably at very late Pliocene (1.35 Myrs BP). The total

hydrocarbon production (oil and gas) from the Eocene shale
is about 275 mg per gm kerogen, with a kerogen
transformation ratio of 0.48.

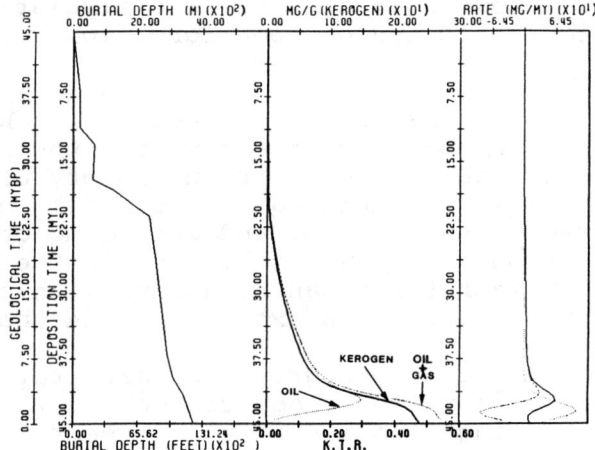

Figure 19. A detailed hydrocarbon generation
history of the Eocene shale (layer 1) in the
COST No. 1 well. Shown are (1) the burial
depth of the layer with time, (2) the
accumulative oil and gas production with
time, and (3) the rate of oil and gas
production. Note that only in the last
8 Ma or so has the accumulative amount
of hydrocarbons exceeded 0.5 mg/gm kerogen.

The early Oligocene shale (layer 3) (Fig. 20), has a
similar hydrocarbon generation history to the Eocene
shale. It enters the "oil window" around middle Pliocene
(4.3 Myrs BP, 3,000 m), and has a peak oil generation at
about the end of the Pliocene (1.35 Myrs BP, 3,500 m). It
still is in its peak gas generation stage with a total
hydrocarbon production (oil and gas) of 154 mg per gm of
kerogen, corresponding to a kerogen transformation ratio
of 0.45.

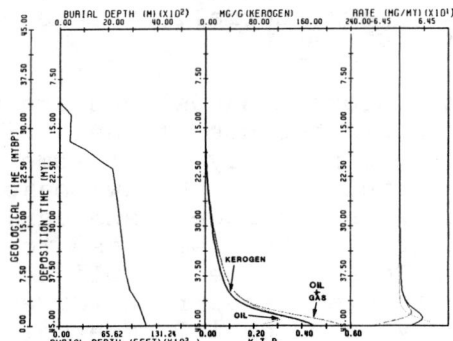

Figure 20. As for figure 19 but for the
early Oligocene shale (layer 3). Note
that significant accumulative production
has occurred only in the last 5 Ma or so.

Layer 5 (late Oligocene mudstone) is just entering
the "oil window" with a total hydrocarbon generation of
84.3 mg per gm of kerogen (78 mg oil and 6.1 mg gas) and
a kerogen transformation ratio of 0.18, but probably
cannot be an effective source rock. The simulation does
not consider biogenetic hydrocarbon formation from
kerogen.

V.   Discussion of Navarin Basin Cost No. 1 Well
Results. From the combined geohistory, the thermal
history and the hydrocarbon generation history of the
Navarin COST No. 1 well simulated by the model consistent
with the observed downhole data, and including other
geological and geophysical data collected in the basin,
the following considerations can be derived, which will be
useful for further hydrocarbon exploration in the basin.

Source Rock. The Eocene gray to dark gray mudstone
and shale (Zone F) is the best source rock in the Navarin
basin with TOC ranging from 1.0 to 2.0 percent. The
Eocene shale reached thermal maturity during the late
Miocene (Fig. 19), arrived at peak oil generation at the
late Pliocene with a total amount of generated
hydrocarbons of about 275 mg/g kerogen. The early
Oligocene shale (Zone E) is next best source rock with TOC
around 1.0%; it matured at the Late Miocene, arrived at
peak oil generation at the end of Pliocene, with a total
amount of hydrocarbons generated of around 154 mg per gram
of kerogen. The late Oligocene mudstone (Layer 3) has
just reached maturity but probably cannot be considered as
an effective source rock in the Navarin COST No. 1 well.
Other younger fine-grained sediments are unlikely to be
good source rocks because they are thermally immature and
have low TOC content (less than 0.5%).
The facts that there are three sub-basins in the
regional scale Navarin Basin, and that the COST No. 1 well
is only located on a structure bench, indicates that it
will be more favorable for the Eocene and early Oligocene
(layers 3 and 5) to generate significant amounts of
hydrocarbons in the sub-basins. The Cretaceous rocks below
the major unconformity are thermally mature enough to be
good source rocks.

Reservoir Rock. Miocene sandstones are the most
promising reservoir rocks because of their high porosity
and permeability. There probably exist some potential
reservoir rocks deposited during the two major uplifts in
the Tertiary (one between the Eocene and early Oligocene,
one in the early Oligocene) because these rocks are
closest to the major Eocene and early Oligocene source
rocks. Also, the late Cretaceous rocks directly below the
major unconformity may be potential reservoir rocks
reflecting a variety of depositional environments, because
(1) they are coarse-grained rocks deposited in streams
and distributary channels and possibly are strongly
influenced by the deltaic distribution system (Turner et
al., 1984); (2) the erosion during the major unconformity

period might lead to both coarse-grained sediments, and secondary and fracture porosity; and (3) the Cretaceous rocks are close to the major source rock above the unconformity, which may provide a good chance for them to receive hydrocarbons generated from the source rock which migrate downwards.

Exploration targets. Based on the geohistory, thermal history and generation history from the model and other geological and geophysical data, exploration for hydrocarbons in the Navarin basin should most likely be targeted at (i) Eocene and Oligocene strata and (ii) Late Cretaceous strata right below the late Cretaceous-early Tertiary unconformity.

Eocene and Oligocene Strata. Eocene and early Oligocene organic rich mudstones and shales (Zones E and F) are thermally mature and are able to generate significant amounts of hydrocarbons. Late Oligocene shelf sands are the best reservoir rocks encountered in the COST No. 1 well. Oligocene and Eocene turbidite deposition may be present along sub-basin margins and as detrital aprons around basement highs (Steffy et al., 1985). Also during the two major uplifts in the Eocene and Oligocene, some reservoir rock-types and favorable facies changes were probably formed in the basin. In addition to these possible stratigraphic traps, seismic surveys indicate the existence of anticlines and growth faults along the flanks of the sub-basins (Marlow, 1979), which means that potential structural traps and growth-fault-related traps may also exist.

Late Cretaceous Strata directly below the unconformity. Eocene and early Oligocene source rocks can not only be the major hydrocarbon sources for late Cretaceous reservoir rocks right below the unconformity, but also they can be effective seals for these reservoir rocks. The late Cretaceous-early Tertiary unconformity may act as a migration path for hydrocarbons generated from the Eocene and early Oligocene source rocks. The possible existence of various stratigraphic traps in the depth range of the late Cretaceous-early Tertiary unconformity, should then be considered as potential candidates for hydrocarbon accumulation.
From this study we conclude that: (1) The dynamic model is a useful tool for hydrocarbon explorationists to make quantitative studies of geohistory, thermal history, hydrocarbon generation and migration history in a sedimentary basin, A 1-D fluid flow model is an appropriate first vehicle to use in a frontier area where only one or a few wells have been drilled, subject to the caveat that the dominant escape of fluid from the sedimentary fill is indeed predominantly vertical.
(2) The major source rocks in the Navarin basin are likely to be Eocene and early Oligocene organic rich mudstone and shales, which are mature enough to generate significant amounts of hydrocarbons.

(3)   The most prospective exploration plays in the Navarin basin are the Eocene and Oligocene strata, and, perhaps the late Cretaceous strata right below the late Cretaceous-early Tertiary unconformity, although the effect of the igneous intrusive on the thermal maturity of sediments in the vicinity of the intrusive has yet to be investigated.

(4)   The two major source rocks have been overpressured since Oligocene time, providing a mechanism for the expulsion of generated hydrocarbons from the source rocks into surrounding reservoir rocks.

VI.   Summary.   The simple recognition that hydrocarbons are produced by thermal degradation of organic material, forces an examination of the interlocked evolution of sedimentary burial and dynamical interaction, of the evolving thermal regime in the sub-surface, and of hydrocarbon generation, migration and accumulation.   These three dominant components are by no means capable of being examined in isolation - each has an impact on the others.

The major concerns of quantitative methods in basin analysis are to examine the inter-relations among the various components, to quantify the dynamical behavior so that basinal processes can be modeled, to use present day data to constrain generic processes and their applications to particular basins, and to provide an assessment of the most likely sites of hydrocarbon accumulations at the present day ahead of expensive drilling.

Such assessments must also provide a measure of their probability of being right based on the best sensitivity studies possible with the data and models to hand.

In the long range thrust to reach that goal, basin analysts address a number of individual problems, with others being less well developed, in order to exhibit as sharply as possible the role of the individual facet under consideration.   But a gradually more integrated picture of interwoven development is being developed as these facets are assimilated into the developing body of knowledge.

Acknowledgments.   The work reported here was supported by the Industrial Associates of the Basin Analysis Group at the University of South Carolina.

Appendix.   1) Geohistory model.   A.   Deposition, Compaction and Fluid Flow.   Compaction of sediment is the result of fluid expulsion caused by the increasing overburden.   The fluid movement in the sediment is described by combining mass conservation and a modified Darcy's equation

$$\frac{\partial}{\partial t}(\rho_f e) = \frac{\partial}{\partial \xi}\left[\frac{k_z \rho_f}{\mu(1+e)}\frac{\partial P}{\partial \xi}\right] \tag{1}$$

where e is the void ratio of the rock, ρ_f is the density of the fluid, μ is the viscosity of the fluid, k_z is the vertical permeability of the rock, and P is the fluid pressure in excess of hydrostatic.   The fully compacted depth coordinate, ξ, represents the net thickness of the rock such that when depositing a new layer of sediment, the previously deposited sediments do not move in ξ. The physical depth z, is related to ξ by $d\xi/dz = (1-\phi)$, where the porosity, ϕ, is related to the void ratio, e, by $e(1-\phi) = \phi$. Darcy's law, expressing flow as a function of permeability and overpressure gradient, can then be applied correctly.

The excess fluid pressure, P, is represented as a function of frame pressure, P_f, assuming no pressure is generated during compaction by, for instance, hydrocarbon liquid or gas production from solid kerogen by

$$p = \int_{\xi}^{\xi^*} g\ (\rho_s - \rho_f)\ d\xi - (P_f - P_{f*}) \tag{2}$$

where g is the acceleration due to gravity, ρ_s the solid density of the rock, P_f is the frame pressure (a prescribed function of void ratio (e) for each lithology), and * denotes the appropriate value on the sediment surface (P = 0 on ξ_*).

All the variables are time-dependent and are related to the sedimentary deposition rate through the increase in z with time. Solving equation (1) determines the changes of sediment thickness, porosity, permeability, pressure and fluid flow rate with time and depth (Cao et al., 1986).

B.   Permeability and Frame Pressure Laws, and Method of Solution. The choice of dependence of permeability and frame pressure on void ratio for various lithologies has had a long and chequered history (Ungerer et al., 1984). Proponents can be found for determining the behavior ab initio, while others note that the complexity and physical processes over geologic time are so variable that quasi-empirical relationships are to be preferred. We have sided with this second viewpoint and , based on measurements from several hundred wells (Gleason, 1983), we have used the following relationships in our calculations (units: cgs):

1. Frame pressure, P_f, is a function of void ratio, e,:

Shale : $P_f = P_{f*}\ (e\ /\ e_*)^{-A}$ \hfill (3)

with the surface value $e_* = 3$ (corresponding to 75% porosity) and the scaling pressure $P_{f*} = 6.9 \times 10^5$ dyne cm^{-2} . The parameter 'A', has a value of around 3.5 for an average of several dozen Gulf Coast wells.

Sand : $e = e_* - C\ (P_f - P_{f*})$ \hfill (4)

where $C \cong 7 \times 10^{-10}$, $e_* \cong 0.6$, $P_{f*} \cong 1.0$ dyne cm^{-2}.

2. Permeability, k, is a function of void ratio

Shale: $k = k_*\ (e\ /\ e_*)^B$, with $k_* \cong 10^{-12}$ cm^{+2}, (5)

Sand: $k = k_*(e / e_*)^B$, with $k_* \cong 3 \times 10^{-7}$ cm^{+2} (6)

Note that the surficial ($e = e_*$) shale permeability is about five orders of magnitude smaller than the sand permeability so that the flow, and the overpressuring, are dominantly controlled by the shale layers.

The two parameters A and B for shale are at our disposal, since small changes in those parameters markedly influence the fluid pressure gradient and permeability.

Since the excess fluid pressure gradient is limited to being less than about 0.5 psi/ft, the dominant control on the dynamical evolution is through the exponent B of the shale permeability. While both A and B appear to be around 4.0, there is sufficient variation in their values from well to well that we have treated them as empirical constants to be determined. This treatment then leads to the method of solution. First "freeze" the parameter A since fluid pressure gradients do not vary by orders of magnitude. Then run the calculation of fluid flow/compaction for a fixed value of B. Check the degree to which the predicted formation thicknesses agree with observed values at the present day. Adjust B until a minimum discordance obtains. (Experience with many wells has shown that with an initial B set at 4.0, the final B will be between about 2.5 to 5.0, and convergence obtains after about 3 iterations). Then "freeze" B at its best value, and let A be allowed to vary until either the predicted porosity with depth or the predicted fluid pressure with depth (but not both) come into minimum discord with the observed values. (Again, experience has shown that an initial A set at 4.0 yields a final A with range 3.0 to 5.0 after about 3 iterations).

The sequence is repeated: A is once more frozen and B adjusted. This second sequence of operations normally adjusts B by about 10%. The re-freezing of the "new" B and subsequent re-determination of A normally varies A by about 10%. Further repetitions of the sequence hardly vary A and B at all.

The permeability and frame pressure variations with void ratio are then determined empirical relations. The burial history then determines the evolution of fluid flow, formation thickness, porosity with depth, fluid pressure with depth, and permeability with depth with no other parameters being varied. The results are as shown in the text.

C. Basement Subsidence. The total basement subsidence is caused by the weight of the sediments on the basement, the weight of the water above the sediment surface and tectonism. The basement subsidence caused by tectonism can be calculated by (Steckler and Watts, 1978):

$$Y = S \left\{ \frac{\rho_m - \rho_s}{\rho_m - \rho_w} \right\} + Wd - \Delta SL \left\{ \frac{\rho_m}{\rho_m - \rho_w} \right\}   (7)$$

where Y is the tectonically caused basement subsidence, S is the sediment thickness, Wd is the water depth at the time of sediment deposition, ΔSL is the change in elevation of mean sea level, ρ_m is the mantle density, ρ_w is the water density and ρ_s is the average sediment density (Steckler and Watts, 1978).

D. Cementation and Dissolution. In the event of cementation or rock dissolution playing major roles we simulate the effects of cementation and dissolution indirectly in terms of the evolution of formation permeability with

$$K = Kp \qquad in \ t < T_1$$

$$K = Kp \left[1.0 - \left\{ \frac{t - T_1}{T_2 - T_1} \right\} C(-D) \right], \ in \ T_2 \geqslant t \geqslant T_1 \qquad (8)$$

$$K = Kp [1.0 - C(-D)], \quad in \ t > T_2$$

where C(D) is the fractional degree of cementation (dissolution), $T_1(T_2)$ is the time at which either cementation or dissolution starts (ends), and Kp is the formation permeability immediately prior to cementation (dissolution) onset.

It can, of course, be argued that this is an extremely primitive method of accommodating for cementation and/or dissolution processes. One might suggest that use of the dependences of mineral concentrations on solubility, temperature, pressure, activity, mobility, etc., would be a much preferred path to follow. While such an approach is, indeed, preferable in principle, from a pragmatic point of view it adds a complexity to the basic problem which is not justified by data availability. The point is that one would have to track mineral concentrations for all times and depths of interest including allowing for variable depositional amounts, saturation in mixed concentrations of varying minerals, ion replacement effects in phases, etc.. Such a program of research would be a major undertaking which is hardly justified by the imprecision of our knowledge of the constitutive equations, or of the input data. We believe that the simple, primitive rule above is a good method of providing an estimate of the dominance, or otherwise, of diagenetic effects on the flow of fluid in sedimentary basins.

E. Fracture Evolution. Fracture evolution is simulated by its effects on formation permeability: For pore pressure, Pp, lower than a critical fraction, f, of the lithologic overburden pressure, PL, fractures are closed, for Pp $\geqslant$ fPL the formation permeability is modeled by

$$Kf = K + Ks * (1.0 - Pc/Pp)^2 \qquad Pp \geqslant fPL$$

$$Kf = K \qquad\qquad\qquad\qquad Pp < fPL \qquad (9)$$

where K is the permeability in the absence fractures, Ks is the permeability of sand, Pp is the total pore fluid pressure (hydrostatic plus excess), and $P_C = fPL$ is a critical pressure (Eaton, 1969; Palcianskas and Domenico, 1980), normally taken to be around 0.8 - 0.9 of the lithology overburden. We use 0.85.

(2) Thermal History Model. A. Heat Flux Temperature. The simplest form to describe heat flux variation with time is the single parameter form

$$Q(t) = Q_o (1 + \beta t) \tag{10}$$

or

$$Q(t) = Q_o \exp(\beta t) \tag{11}$$

where Q(t) is the basement heat flow as a function of time, Q_o is the present day heat flow, t is time in million years and β is a constant to be determined by minimizing the difference in predicted versus measured vitrinite reflectance with depth (see later). The formation temperature at time t and depth z(t) below the sediment-water interface is

$$T(t, z(t)) = T_s + Q(t) \int_0^{z(t)} dz'/K(z') \tag{12}$$

where T_s is the surface temperature and K(z) is the sedimentary thermal conductivity at depth z(t) described through

$$K(z) = (K_{fl})^{\phi(z)} (K_s)^{(1-\phi(z))} \tag{13}$$

where K_{fl} is the thermal conductivity of the pore fluid, K_s is the thermal conductivity of the solid rock and $\phi(z)$ is the porosity at depth z (Lewis and Rose, 1970; Andrews-Speed et al., 1984). Numerical values are lithology dependent and are also provided in the above two references.

B. Thermal Maturity of Source Rocks. Vitrinite reflectance changes with time and depth satisfy (Lerche et al., 1984)

$$R(t) = \left[R_s^{\frac{1}{2}} + I(t) \left\{ \sum_{j=1}^{n} (R(z_j)^{\frac{1}{2}} - R_s^{\frac{1}{2}} \right\} / \sum_{j=1}^{n} I(j) \right]^2 \tag{14}$$

where R(t) is the formation vitrinite reflectance at time t when the formation is at depth z(t) post deposition, I(j) is a time-temperature integral (characterizing the maturity of a vitrinite sample deposited at time t_j, presently at depth z_j) with

$$I(j) = \int_{-tj}^{0} \exp \left\{ \frac{T - T_c}{T_D} \right\} dt \quad , \quad \text{in } T \geqslant T_c \tag{15}$$

with the integrand zero in $T < T_c$, where T is the temperature along the burial depth, T_c and T_D are fixed at 295°K and 200°K, respectively (Toth et al., 1981; Lerche et al., 1984), $I(t)$ is $I(j)$ at time t, and $R(z_j)$ is the measured vitrinite reflectance at depth z_j. R_s is the value of vitrinite reflectance at the depositional surface, taken to be 0.2.

Waples' Time-Temperature Index (TTI) is defined by (Waples, 1980)

$$TTI = \sum_{nmin}^{nmax} (\Delta T_n)(r^n) \tag{16}$$

where t_n is the length of time spent by the sediment in the temperature interval n, r is the temperature factor, n is the appropriate index value, nmax (nmin) is the n-value of the highest (lowest) temperature interval encountered (Waples, 1980). In Waples (1980) a constant thermal gradient was used to calculate temperature change with time and depth. Extension to a variable (in time and depth) thermal gradient, dependent on both heat flow, $Q(t)$, and thermal conductivity, $K(z)$, is used in the present model.

(3) Hydrocarbon Generation Model. A. Tissot's model. Hydrocarbon generation is simulated by

$$-\frac{dX_i}{dt} = K_{1i} X_i,$$

$$\frac{dU_j}{dt} = K_{2j} \tag{17}$$

$$Y = \sum_i Y_i$$

$$\sum_i X_{io} + \sum_i Y_{io} + \sum_j U_{jo} = \sum_i X_i + \sum_i Y_i + \sum_j U_j$$

with K_{1i} and K_{2j} given by

$$K_{1i} = A_{1i} \times \exp(-E_{1j}/RT)$$

$$\tag{18}$$

$$K_{2j} = A_{2j} \times \exp(-E_{2j}/RT)$$

where X_i is the concentration of component i in Kerogen, Y is the product (oil) from kerogen degradation, U is the product (gas) from oil cracking, K_{1j} is the reaction constant for gas cracked from oil, A_{1i} and A_{2j} are the frequency factors, E_{1i} and E_{2j} are the activation energies, R is the gas constant, T is the temperature and t is the time. (Tissot and Welte, 1978).

B. A Modified Model. Hydrocarbon generation is simulated by

(a) First stage - Kerogen Degradation

(i) Total Kerogen degradation for type n

$$\frac{dX_n}{dt} = - X_n \left(\sum_{i=1}^{6} K_{1i} + \sum_{j=1}^{6} K_{2j} \right) \tag{19}$$

where X_n is the concentration of Kerogen, K_{1i} is the reaction constant for oil with bond type i and the K_{2j} is the reaction constant for different gases (CH_4, H_2, N_2, etc.).

(ii) Direct gas production

$$\frac{dU_j}{dt} = K_{2j} \times X_n \tag{20}$$

(iii) Direct oil production

$$\frac{dY_i}{dt} = K_{1i} \times X_n - Y_i \sum_{j=1}^{6} K_{2j} \tag{21}$$

The amount of oil type i is Y_i, total oil is $\sum_{i=1}^{6} Y_i$.

(b) Second Stage - Oil Cracking. Gas type j from oil cracking:

$$\frac{dU'_j}{dt} = K_{2j} \sum_{i=1}^{6} Y_i \tag{22}$$

Total gas is $G = \sum_{j=1}^{6} G_j$, with gas of type i being $G_j = U_j + U'_j$. The value of K_{1i} in this model is as in Tissot's model as is the value for $k(CH_4)$, The values of K_{2j}, the reaction constants for gas production, are poorly known. If the reaction constants of different gaseous compounds are associated with the relevant molecular weights (the lighter a compound the less energy it needs to react and so the higher its reaction constant), then the reaction constant is inversely proportional to a compound's molecular weight, e.g. $K(CO_2) = (16/44) K(CH_4)$. We have used this formulation here in lack of anything better being available.

REFERENCES

1. Airy, G.B., 1855, On the computation of the effect of the attraction of mountain-masses as disturbing the apparent astronomical latitude of station of geodetic surveys: Philosophical Transactions of the Royal Society of London, vol. 145, p. 101-104.

2. Andrew-Speed, C. P., E. R. Oxburgh and B. A. Cooper, 1984, Temperature and depth-dependent heat flow in West North Sea, AAPG Bull., v. 68, p. 1764-1781.

3. Brooks, J. and Welte, D., 1984, Advances in Petroleum Geochemistry, vol. 1, 333 pps., Academic Press, New York.

4. Cao, S., W. H. Glezen, and I. Lerche, 1986, Fluid flow, hydrocarbon generation and migration: A quantitative model of dynamical evolution in sedimentary basins, Proceedings of The Offshore Technology Conference (Houston, TX) paper 5182, vol. 2, p. 267-276.

5. Dahlberg, 1982, Applied Hydrodynamics in Petroleum Exploration: New York, Springer-Verlag, 171 p.

6. Eaton, B. A., 1969, Fracture gradient prediction and its application in oil field operation: Journal of Petroleum Technology, v. 21, p. 1353-1360.

7. Falvey, D.A. and M.F. Middleton, 1981, Passive continental margins: Evidence for a pre-breakup deep crustal metamorphic subsidence mechanism: 26th International Geological Congress, Colloque C3.3, Geology of Continental Margins, Supplement to vol. 4, p. 103-114.

8. Guidish, T.M., C.G.St.C. Kendall, I. Lerche, D. Toth, and R.F. Yarzab, 1985, Basin evolution using burial history calculation: an overview, AAPG Bull., v. 69, p. 92-105.

9. Hubbert, 1940, The Theory of Ground-Water Motion, J. Geol. Vol. 48, 785-944.

10. Hunt, 1979, Petroleum Geochemistry and Geology. W.H. Freeman & Co., San Francisco, CA, 617 pps.

11. Lerche, I. and O'Brien, J.J., 1987, Dynamical Geology of Salt and Related Structures, 832 pps., Academic Press, Orlando.

12. Lerche, I., R.F. Yarzab, and C.G.St.C. Kendall, 1984, Determination of paleoheat flux from vitrinite reflectance data, AAPG Bull., v. 68, p. 1704-1717.

13. Lewis, C.R. and SC. Rose, 1970, A Theory relating high temperatures and overpressures. J. Petrol. Technol. 22, 11- 16.

14.   Lopatin, N.V., 1971, Temperature and geologic time as factors in coalification: Akademiya Nauk SSSR Izvestiya Seriya Geologicheskaya, n. 3, p. 95-106.

15.   Marlow, M.S., 1979, Hydrocarbon prospects in Navarin basin province, northwest Bering Sea shelf: Oil and Gas Journal, Oct. 29, 1979, p. 190-196.

16.   McKenzie, D., 1978, Some remarks on the development of sedimentary basins: Earth and Planetary Science Letters, vol. 40, p. 25-32.

17.   Milliman, 1974, Marine Geology, Part 1, Springer-Verlag, New York.

18.   Palcianskas, V.V. and P. A. Domenico, 1980, Microfracture development in compacting sediments: A relation to hydrocarbon maturation kinetics: AAPG   Bull., v. 64, p. 927-937.

19.   Sclater, T.G. and P.A.F. Christie, 1980, Continental stretching: an exploration of the post-mid-Cretaceous subsidence of the Centrtal North Sea basin: Journal of Geophys. Res., v. 85, No. 67, p. 3711-3739.

20.   Steckler, M.S. and A. B. Watts, 1978, Subsidence of the Atlantic-type continental margin off New York: Earth and Planetary Science Letters, 41 (1978), p. 1-13.

21.   Steffy, D.A., R.F. Turner, G.C. Martin, and T.O. Flett, 1985, Evolution and petroleum geology of the Navarin basin, Bering Sea, Alaska: Oil and Gas Journal, Aug. 5, 1985, p. 116-124.

22.   Tissot, B. and D.H. Welte, 1978, Petroleum formation and Occurrence, New York, Springer-Verlag, p. 538.

23.   Tissot, B., and Espitalie, J., 1975, L'evolution thermique de la matiere organique des sediments: applications d'une simulation mathematique: Revue Insitut Francais du Petrole, vol. 30, p. 743-777.

24.   Tissot, B., 1969, Premieres donnees sur les mecanismes et la cinetique de la formation du petrole dans les sediments: simulation d'un schema reactionnel sur ordinateur: Revue Institut Francais du Petrole, vol. 24, p. 470-501.

25.   Toth, D.J., I. Lerche, D.E. Petroy, R.J. Heyer, and C.G.St.C. Kendall, 1981, Vitrinite reflectance and the derivation of heat flow changes with time: in Advances in Organic Geochemistry, 1981, New York, John Wiley and Sons, p. 588-596.

26.   Turner, R.F., C.M. McCarthy, D.A. Steffy, M.B. Lynch, G.C. Martin, K.W. Sherwood, T.O. Flett, and A.J. Adams, 1984, Geological and operational summary, Navarin Basin COST No. 1 well, Bering Sea, Alaska. OCS Report MMS 84-0031. Anchorage: U.S. Department of the Interior, Mineral Management Service.

27.   Ungerer, P. et al., 1984, Geological and geochemical methods in oil exploration, principles and practical examples, AAPG Memoir 35, p. 53-77.

28.   Waples, D.W., 1988, Workshop on Quantitative Dynamic Stratigraphy, Abstract entitled "Maturity Modeling of Sedimentary Basins: Approaches, Limitations, and Assessment of Future Developments" p. 19, Denver, Co.

29.   Waples, P.W., 1980, Time and temperature in petroleum formation: application of Loptin's method to petroleum exploration: AAPG Bull., v. 64, p. 916-926.

30.   Welte, D.H. and M. A. Yukler, 1981, Petroleum origin and accumulation in basin evolution - a quantitative model, AAPG Bull., v. 65, p. 1387-1396.

Table 1. Basic Input data for Navarin Basin COST No. 1 Well

Layer No.	Depth(1) at base of form. (ft)	Age at base (MYBP)	Lithology	Paleowater depth (ft)	Kerogen content % Type I	Type II	Type III
13	1920.0	3.4	mudstone	432.0	–	–	–
12	3180.0	5.3	sandy mudstone	300.0	–	–	–
11	3860.0	7.9	fine-grain muddy sandstone	600.0	–	–	–
10	5010.0	18.6	"	600.0	0.0	0.4	0.6
9	5360.0	21.2	"	600.0	0.0	0.4	0.6
8	5704.0	23.7	"	600.0	0.0	0.4	0.6
7	7130.0	24.8	"	300.0	0.0	0.4	0.6
6	9450.0	26.7	sandy mudstone	600.0	0.0	0.4	0.6
5	11100.0	28.0	mudstone	1500.0	0.0	0.4	0.6
4	11100.0	32.0	hiatus	0.0	0.0	0.0	0.0
3	12280.0	34.0	mudstone	1500.0	0.0	0.5	0.5
2	12280.0	38.0	hiatus	0.0	0.0	0.0	0.0
1	12780.0	45.0	mudstone	600.0	0.0	0.7	0.30

Notes: (1) Depths are measured from the Kelly Bushing, which was 85 feet above mean sea level.

(2) Age is based on GSA 1983 Geologic Time Scale.

(3) Lithology and paleowater depth and kerogen content are based on OCS Report MMS 84–0031(Turner et al., 1984).

Table 2.   Thickness change (m) of each modeled layer with time (My).

Deposition time My	Layer Number	1 shale	3 shale	5 shale	6 sandy mud	7 sandy	8 sandy	9 sandy	10 sandy	11 sandy	12 gray shale	13 mudstone	Total Thickness
7	(1)	230.0											
11	(2)	229.4											
13	(3)	196.8	477.2										
17	(4)	187.5	436.8										
18.3	(5)	181.0	401.1	691.5									
20.2	(6)	171.3	385.0	558.1	800.1								
21.3	(7)	164.4	373.4	530.1	734.1	553.5							
23.8	(8)	161.3	368.3	528.2	729.8	514.6	157.1						
26.4	(9)	159.1	364.8	524.6	726.8	508.7	123.3	154.0					
37.1	(10)	153.3	357.0	515.5	717.5	487.1	118.3	115.8	426.8				
39.7	(11)	151.4	355.9	513.6	712.4	476.0	115.5	113.1	380.5	261.5			
41.6	(12)	148.0	355.1	511.3	705.0	452.9	109.6	107.3	361.8	211.3	440.5		
45.00	(13)	144.6	353.4	508.7	701.8	442.2	107.0	104.8	353.6	206.7	375.0	431.2	3729.0 Predicted
Measured thickness		152.4	359.7	502.9	707.1	434.6	104.8	106.7	350.5	207.3	384.1	427.6	3737.7 Measured

Note: Layers 2 and 4 are unconformities treated as depositional hiatae.

Table 3. Comparison of Modeled porosity with measured porosity

Layer No.	1	3	5	6	7	8	9	10	11	12	13
Porosity %	shale	shale	shale	sandy shale	muddy sand	muddy sand	muddy sand	muddy sand	muddy sand	sandy mud	mud
Modeled Average Porosity	12.1	12.4	12.7	21.9	29.2	32.0	33.1	35.1	37.1	28.2	36.5
Measured Average Porosity	12.9	11.5	15.2	18.7	37.0	31.5	30.2	31.0	35.1	–	–

CHAPTER 9

An Anayltical Solution to a Pumped Leaky Aquifer System

T. A. Rizk*
J. M. Bownds**
M. M. Stevens**

ABSTRACT. The general analytical solution to a pumped axisymmetric, homogeneous, anisotropic aquifer, with leaky confining layers is obtained here. This solution is useful in examining the various mechanisms controlling ground water flow in pumped leaky aquifer systems. Specifically, the analytical result can be used as a bench mark for numerical analysis. It also represents a basis from which hydraulic parameters are obtainable. The structural elements of the solution are readily generalized to multi—layered aquifer systems. The analytical solution offered here is for a coupled leaky aquifer aquitard system and the solution is validated with a numerical finite element solver and it is examined in light of the classical solutions to the problem.

1. INTRODUCTION. The problem of pumped aquifers have been of interest since the 1930's. Theis [1] considered the problem of a one dimensional radial fluid flow to a well. Hantush and Jacob [2] and Hantush [3,4] considered the problem of a confined aquifer with a partially penetrating pump where the assumption of strictly vertical flow in the aquitard with negligible storage effects, the leakage across the interface is replaced by a volumetric source term in the aquifer domain. This equivalent source term is found to be proportional to the aquifer drawdown. The work of Neuman and Witherspoon [5] considered an axially symmetric aquifer system having leakage from the confining layers in which the fluid flow is vertical, and a pumped aquifer in which the fluid flow is radial. More recently, Chen et. al. [6] studied a closely related problem to the Hantush type problem showing that the aquitard flow may have radial velocity components, contrary to the assumptions in the work of Hantush and Jacob, Herrera [7], Frind [8], and Neuman and Witherspoon. It is also shown by Chen, that for some cases, the analyses produce qualitative results that are very similar. Differences occur, however, when the aquitard is anisotropic and the radial hydraulic conductivity is significant when compared with the vertical hydraulic conductivity.

* TVA Engineering laboratory, Norris, Tennessee.
** Oak Ridge National Laboratory, Oak Ridge, Tennessee.

The model developed for this paper differs from those above in that the leakage from the confining layers is characterized as a boundary effect at the interface between the pumped aquifer and the confining layers to yield a boundary–coupled, inhomogeneous system. The system solution is obtained and validated with a finite element simulation.

2. PROBLEM DESCRIPTION. The mathematical description of drawdown in a confined pumped axially symmetric aquifer occupying a cylindrical domain,

(2.1) $\mathbb{D} = \{ (r,z,t) : 0<r<+\infty , 0<z<b , 0<t \},$

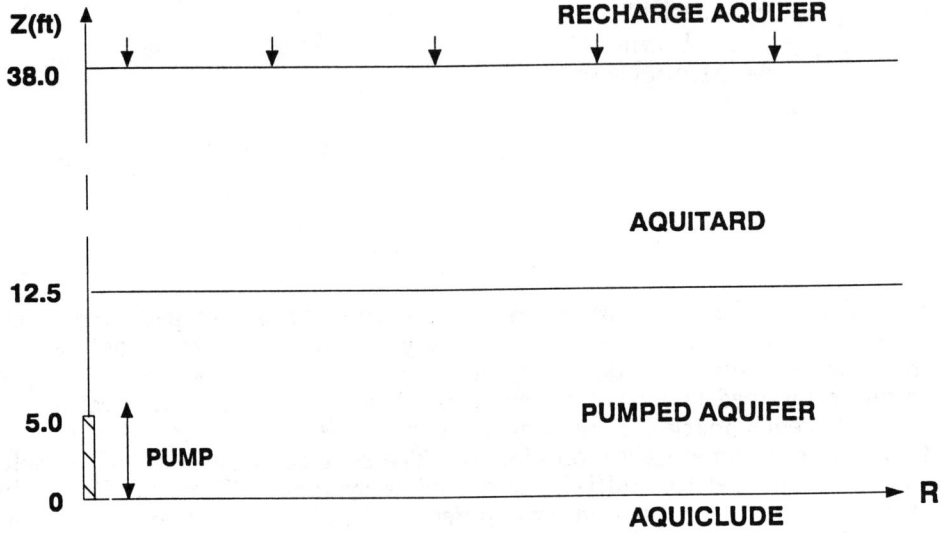

Figure 1. Schematic Representation of a Two–layered Aquifer Model

as shown in cross–section in figure 1. The problem is referred to as the artesian well problem by Hantush [3]. As shown in Freeze and Cherry [9], the Darcian equation of drawdown to the artesian well problem is of the form

(2.2a) $\dfrac{1}{r} \dfrac{\partial}{\partial r} \left[r \dfrac{\partial s}{\partial r} \right] + \kappa \dfrac{\partial^2 s}{\partial z^2} = \dfrac{1}{\nu} \dfrac{\partial s}{\partial t},$

where

(2.2b) $\kappa = \dfrac{k_z}{k_r} \quad \text{and} \quad \nu = \dfrac{k_r}{S_s}.$

The parameters k_r and k_z are radial and vertical conductivities and S_s is the aquifer specific storage. The customary pumped aquifer initial condition is that the initial head is given by the equilibrium piezometric surface. This means that the drawdown satisfies the initial condition

(2.3) $s(r,z,0) = 0.$

The Dirichlet boundary condition at infinity is,

(2.4) $s(\infty,z,t) = 0,$

and the aquifer boundary condition at the origin is of the form

(2.5) $\lim_{r \to 0} r \dfrac{\partial s}{\partial r} \begin{cases} -q & d \le z \le \ell \\ 0 & \text{otherwise} \end{cases},$

with $q = \dfrac{Q}{2\pi K_r (\ell - d)}$;   Q is the volumetric pump rate and b is the aquifer thickness.

At the upper and lower boundaries

(2.6) $\dfrac{\partial s}{\partial z}(r,b,t) = -\varphi(r,t)$ at $z = b$.

and

(2.7) $\dfrac{\partial s}{\partial z}(r,0,t) = 0.0$ at $z = o,$

where $\varphi(r,t)$ is a positive real valued function.

The general procedure to solve the related homogeneous problem (both aquifer confining layers are impermeable) has been outlined in Hantush [3]. Difficulties have arisen in attempting to describe the contributions of the inhomogeneities at the boundaries (leakage from confining layers). In the ensuing discussion, a brief history of the various attempts to solve the problem is presented, and the exact analytical solution to the problem outlined in (2.1) through (2.7) is developed. Finally, the classical application to a leaky aquifer aquitard system is obtained.

3. HISTORICAL DEVELOPMENT. Theis[1] examined the problem of radial fluid flow to a fully penetrating well. The governing system equation is

(3.1) $\dfrac{1}{r} \dfrac{\partial}{\partial r} \left[r \dfrac{\partial s}{\partial r} \right] = \dfrac{1}{\nu} \dfrac{\partial s}{\partial t}$

where

$\nu = \dfrac{k}{S_s},$

with the conditions

(3.2) $s(r,z,0) = 0$

(3.3) $s(\infty,z,t) = 0$,

and

(3.4) $\lim\limits_{r \to 0} r \dfrac{\partial s}{\partial r} = q$,

with $q = \dfrac{Q}{2\pi k b}$.

The problem (3.1) through (3.4) has the solution

(3.5) $T(r,t) = q\, W\!\left[\dfrac{r^2}{4\,\nu\, t}\right]$,

where $T(r,t)$ is defined here as the Theis drawdown, and $W(u)$ is the exponential integral, commonly known among the ground water community as the Theis Well function. The properties of the function $W(u)$ are listed in Abramowitz and Stegun [10].

Considering the problem of a pumped anisotropic aquifer with impermeable confining boundaries, the governing equations are a modified version of (2.2) through (2.7) except that $\varphi(r,t)$ is zero. The problem has the solution

(3.6) $H(r,z,t) = -\dfrac{q}{b} \displaystyle\sum_{n=0}^{\infty}{}' \; a_n\, W\!\left\{\dfrac{r^2}{4\nu t}, \dfrac{n\pi r}{b}\sqrt{\kappa}\right\} \cdot \cos\!\left[\dfrac{n\pi z}{b}\right]$,

where $H(r,z,t)$ is the drawdown in an impermeable aquifer with a partially penetrating pump. The function $W(u,\beta)$ in equation (3.6) referred to as the Hantush well function; it is an extension of the Theis well function described previously. The series coefficients a_n are

(3.7) $a_n = \dfrac{\sin\left[\dfrac{n\pi \ell}{b}\right] - \sin\left[\dfrac{n\pi d}{b}\right]}{n\pi/b}$.

The full solution for a leaky pumped aquifer is considered next.

4. PROBLEM SOLUTION. Following the procedure outlined in Bownds [11, 12,13], the governing aquifer drawdown equation is Laplace transformed in time, and Fourier transformed in vertical space, to yield a singular Sturm–Liouville problem in radial extent. The resulting equation is solved and then inverted back to real space and time. The structure of the resulting Green's function and a useful representation of its inverse transform are discussed below.

4.1. Aquifer Two–Dimensional Solution. The system governing equation (2.2a) is Laplace transformed with respect to time, using p as the transform variable. This is followed by a finite cosine transformation on [0,b]. The use of this particular transform in space is, of course, dictated by the boundary conditions. The resulting drawdown equation is of the form

(4.1)
$$\frac{1}{r}\frac{d}{dr}\left[r\frac{d\tilde{s}}{dr}\right] - \left[\frac{p}{\nu} + \kappa\left[\frac{n\pi}{b}\right]^2\right]\tilde{s} = (-1)^n \kappa\tilde{\varphi}(r,p),$$

where we define

(4.2)
$$\tilde{\varphi} = \mathcal{L}(\varphi),$$

with $\mathcal{L}$ denoting the Laplace transformation. The resulting transformed boundary condition (2.5) is

(4.3)
$$\lim_{r\to 0} r\frac{d\tilde{s}}{dr} = -\frac{q}{p}\left[\frac{\sin\left[\frac{n\pi\ell}{b}\right] - \sin\left[\frac{n\pi d}{b}\right]}{n\pi/b}\right],$$

and the transformed equation (2.4) is

(4.4)
$$\lim_{r\to\infty} \tilde{s}(r,n,p) = 0,$$

where $n = 0,1,2,\ldots$.
The system of equations (4.1) through (4.4) has the solution

(4.5)
$$\tilde{s}(r,n,p) = A_{np} K_0(\omega_n r) + B_{np} I_0(\omega_n r) + P(r,n,p),$$

where $P(r,n,p)$ is a particular solution of the inhomogeneous, doubly transformed equation (3.1). The functions K_0 and I_0 are the usual modified Bessel functions, and the quantity ω_n is given by

(4.6)
$$\omega_n = \left[\frac{p}{\nu} + \kappa\left[\frac{n\pi}{b}\right]^2\right]^{1/2}$$

As shown in Bownds and Rizk [14] and Rizk [15], there exists a Green's Function for this boundary value problem, defined for all complex ω with $\mathrm{Re}(\omega) > 0$, as

(4.7)
$$G(\omega_n r, \omega_n \rho) = \begin{cases} \rho\, K_0(\omega_n r)\, I_0(\omega_n \rho) & \rho \le r \\ \rho\, I_0(\omega_n r)\, K_0(\omega_n \rho) & \rho > r \end{cases},$$

and the general solution to the transformed boundary value problem (4.1) through (4.4) is of the form

$$(4.8) \qquad \tilde{s}(r,n,p) = \frac{qa_n}{p} K_0(\omega_n r) - \kappa(-1)^n \int_0^\infty G(\omega_n r, \omega_n \rho) \tilde{\varphi}(\rho,p) d\rho$$

with a_n defined in (4.3). The inversion with respect to the Fourier Transform is

$$(4.9) \qquad \tilde{s}(r,z,p) = \frac{2}{b} \sum_{n=0}^\infty {}' \tilde{s}(r,n,p) \operatorname{Cos}\left[\frac{n\pi z}{b}\right],$$

yields

$$(4.10) \qquad \tilde{s}(r,z,p) = \frac{2}{b} \sum_{n=0}^\infty {}' \left\{ \frac{qa_n}{p} K_0(\omega_n r) \right.$$

$$\left. - \kappa(-1)^n \int_0^\infty G(\omega_n r, \omega_n \rho) \left[(-1)^n \tilde{\varphi}(\rho,p)\right] d\rho \right\} \operatorname{Cos}\left[\frac{n\pi z}{b}\right],$$

where $\sum'$ denotes 1/2 the first term in the summation series. The inversion with respect to Laplace transform yields the two dimensional transient drawdown solution

$$(4.11) \qquad s(r,z,t) = \frac{2}{b} \sum_{n=0}^\infty {}' \left\{ q\, a_n\, W\left[\frac{r^2}{4\nu t}, \frac{n\pi r}{b}\sqrt{\kappa}\right] \right.$$

$$\left. - \kappa(-1)^n \, \mathcal{L}^{-1}\left[\int_0^\infty G(\omega_n r, \omega_n \rho) \, \tilde{\varphi}(\rho,p) \, d\rho\right] \right\} \cdot \operatorname{Cos}\left[\frac{n\pi z}{b}\right].$$

Here use was made of the Laplace transform inversion relation

$$(4.12) \qquad \mathcal{L}^{-1}\left\{ \frac{1}{p} K_0\left[A \sqrt{p + B^2} \right] \right\} = W\left[\frac{A^2}{4t}, B\right].$$

The inverse Laplace transform of the Green's function will be treated next.

4.2. Properties of the Green's Function. In the composite transform domain, the Green's function is

$$(4.13) \quad G(\omega_n r,\omega_n \rho) = \begin{cases} \rho\, K_0(\omega_n r)\, I_0(\omega_n \rho), & r \geq \rho \\ \rho\, K_0(\omega_n \rho)\, I_0(\omega_n r), & r < \rho \end{cases},$$

for any complex number ω_n satisfying

$$(4.14) \quad \left| \arg \omega_n \right| < \pi.$$

The function G has the following properties:

- The Bromwich integral is *real* and may be expressed with a *real* integration variable.

- The Inverse Laplace Transform of G is

$$(4.15) \quad \mathcal{L}^{-1}\left\{ G(\omega_n r,\omega_n \rho) \right\} = E_n(r,\rho,t) = \frac{\rho}{2t} \exp[-\kappa t(n\pi/b)^2].$$

$$\begin{cases} \sum_{k=0}^{\infty} \left[\sum_{\ell=0}^{k} \frac{(-1)^\ell\, u^{k-\ell}}{\ell!\, [\,(k-\ell)!\,]^2} \right] \frac{e^{-u}\, v^k}{4^k} & v \leq u \\ \\ \sum_{k=0}^{\infty} \left[\sum_{\ell=0}^{k} \frac{(-1)^\ell\, v^{k-\ell}}{\ell!\, [\,(k-\ell)!\,]^2} \right] \frac{e^{-v}\, u^k}{4^k} & v > u \end{cases}$$

where $u = \dfrac{r^2}{4\nu t}$, $v = \dfrac{\rho^2}{4\nu t}$.

- In the doubly transformed domain,

$$(4.16) \quad \int_0^\infty G(\omega_n r,\omega_n \rho)\, d\rho = \frac{1}{\omega_n^2} \quad (4.16)$$

for any $\omega_n > 0$.

- Curiously, the Green's function satisfies the numerical inequality

$$(4.17) \quad 0 \leq G(\omega_n r,\omega_n \rho) \leq \frac{0.53336}{\omega_n}.$$

- In real time, the Exponential Integral is bounnded by

(4.18) $0 \leq E_n(r,\rho,t) \leq \frac{\rho}{2t} \exp[-\kappa t(n\pi/b)^2],$

for all $t,r,\rho.$

4.3. Depth Averaging. In some field experiments, the drawdown measurements are taken over the aquifer thickness. Consequently, it is useful to compute the depth averaged drawdown by integrating relation (4.10) with respect to aquifer depth to obtain

(4.19) $\displaystyle \tilde{\bar{s}}(r,p) = \frac{1}{b} \int_0^b \frac{2}{b} \sum_{n=0}^{\infty} {}' \tilde{s}(r,n,p) \cdot \cos\left[\frac{n\pi z}{b}\right] dz.$

The resulting depth averaged aquifer drawdown is of the form

(4.20) $\displaystyle \tilde{\bar{s}}(r,p) = \frac{q(\ell-d)}{bp} K_0(\omega_0 r) - \frac{\kappa}{b} \int_0^{\infty} G(\omega_0 r, \omega_0 \rho) \tilde{\varphi}(\rho,p) d\rho$

where $\omega_0 = \sqrt{p/\nu}$; the pump discharge q is as defined in section 2, and the Green's function is as defined previously.

The first term in relation (4.20) is the depth averaged equivalent Theis radial drawdown. The next term is the correction to the drawdown due to the flux from the confining layer. The depth averaged solution is valid for any boundary flux functional description satisfying typical regularity conditions such as piecewise continuity and integrability on $[0,\infty)$. The depth averaged solution may be used for examining flow characteristics (Rizk, [15]). However, in the application to contaminant transport problems, the vertical variations in flow should be considered, meaning that the variable n in (4.19) would be retained.

5. TREATMENT OF THE BOUNDARY CONDITIONS. Using transform methods described above, a system of integral expressions for the transformed drawdowns in the aquifer and a corresponding adjacent aquitard may be derived with coupling at their common interface. An immediate application would be the aquifer aquitard system described in Hantush [3]. The overlying aquitard is assumed to be an ideal axisymmetric aquitard, with an upper recharge layer. The aquitard fluid flow is solely in response to the pump action in the aquifer. The model domain is shown in Figure 1. The solution procedure of the aquitard governing system of equations of drawdown is similar to the procedure applied to the aquifer.

6. AQUITARD DRAWDOWN SOLUTION. The solution procedure for the aquitard in the coupled leaky aquifer system is similar to the generalized solution approach applied to the pumped aquifer in section 4. The essential differences are in that for the case of the aquitard, the finite Fourier sine transform is applied, and, since there is no pumping in the aquitard, the first term in the homogeneous solution is zero. The second term in the homogeneous solution is coupled to the particular solution to yield a Fredholm integral equation as in the case of the aquifer. The Green's function is uniquely dependent on the aquitard parameters, and the solution proposed in relation (4.11) is directly applicable to any layered aquifer system with an arbitrary number of layers.

6.1. Aquitard governing system of equations. The governing equation for the overlying aquitard is

$$(6.1a) \qquad \frac{1}{r}\frac{\partial}{\partial r}\left[r\frac{\partial s'}{\partial r}\right] + \kappa'\frac{\partial^2 s'}{\partial z^2} = \frac{1}{\nu'}\frac{\partial s'}{\partial t},$$

where

$$(6.1b) \qquad \kappa' = \frac{k'_z}{k'_r} \text{ and } \nu' = \frac{k'_r}{S'_s},$$

with the initial and boundary conditions

$$(6.2) \qquad s'(r,z,0) = 0,$$

$$(6.3) \qquad s'(r,b',t) = 0,$$

and

$$(6.4) \qquad s'(\infty,z,t) \to 0.$$

The term $b'-b$ is the aquitard thickness. The interface matching conditions between the overlying aquitard and the pumped aquifer are continuity of drawdown and flux, that is

$$(6.5) \qquad s'(r,b,t) = s(r,b,t)$$

and

$$(6.6) \qquad -k'_z\frac{\partial s'}{\partial z}(r,b,t) = k_z\frac{\partial s}{\partial z}(r,b,t) = -k_z\varphi(r,t),$$

where the negative sign is due to the conjugate matching at the interface $z=b$. Both of the conditions (6.5) and (6.6) must be satisfied everywhere on the interface.

6.2. An Extension Of Equations For Drawdown. The finite Fourier sine transform, along with the Laplace transform are applied to the drawdown problem (6.1) through (6.6). The resulting drawdown in the aquitard is of the general form

$$(6.7) \qquad \tilde{s}'(r,n,p) = C_{np}K_0(\omega'_n r) + D_{np}I_0(\omega'_n r) + P'(r,n,p).$$

As before, the function P' is a particular solution of the inhomogeneous, doubly transformed aquitard problem. Similar to the solution in section 4,

$$(6.8) \qquad \omega'_n = \left[\frac{p}{\nu'} + \left[\frac{n\pi}{b'-b}\right]^2\kappa'\right]^{1/2},$$

and the Green's function for this boundary value problem, defined for all complex ω' with $\text{Re}(\omega') > 0$, is, as before,

$$(6.9) \qquad G(\omega'_n r, \omega'_n \rho) = \begin{cases} \rho \, K_0(\omega'_n r) \, I_0(\omega'_n \rho) \; , \rho < r \\ \rho \, I_0(\omega'_n r) \, K_0(\omega'_n \rho) \; , r \leq \rho \end{cases}$$

It follows that

$$(6.10) \qquad \tilde{s}'(r,n,p) = C_{np} \, K_0(\omega'_n r)$$

$$+ \frac{\kappa' \, n \, \pi}{b' - b} \int_0^\infty G(\omega'_n r, \omega'_n \rho) \, \tilde{s}'(\rho,b,p) \, d\rho.$$

The determination of C_{np} is accomplished by using the assumption of no pump in the aquitard. Note that

$$(6.11) \qquad r \frac{d\tilde{s}'}{dr}(r,n,p) = - C_{np} \, \omega'_n r \, K_1(\omega'_n r)$$

$$+ r \frac{d}{dr} \int_0^\infty G(\omega'_n r, \omega'_n \rho) \, \tilde{s}'(\rho,b,p) \, d\rho,$$

and as r approaches the origin, we deduce

$$(6.12) \qquad C_{np} = 0$$

from the following Bessel Function property :

$$(6.13) \qquad z K_1(z) \to 1 \quad \text{as} \quad z \to 0.$$

7. COUPLED SOLUTION. Combining the above results, the doubly transformed drawdowns are given by

$$(7.1) \qquad \tilde{s}(r,n,p) = \frac{q}{p} \, a_n \, K_0(\omega_n r)$$

$$+ (-1)^n \kappa \int_0^\infty G(\omega_n r, \omega_n \rho) \frac{\partial \tilde{s}}{\partial z}(\rho,b,p) d\rho,$$

and

$$(7.2) \qquad \tilde{s}'(r,n,p) = \frac{\kappa' \, n \, \pi}{b' - b} \int_0^\infty G(\omega'_n r, \omega'_n \rho) \, \tilde{s}'(\rho,b,p) \, d\rho.$$

When inverted back into real space and time, the two remaining matching conditions,

$$(7.3) \qquad s'(r,b,t) = s(r,b,t)$$

and

(7.4) $-k_z' \dfrac{\partial s'}{\partial z}(r,b,t) = k_z \dfrac{\partial s}{\partial z}(r,b,t) = -k_z \varphi(r,t),$

are added to the system of equations (7.1) and (7.2), and a complete system of boundary integral equations is obtained.

The conjugate boundary condition at $z = b$ is solved using a successive approximations technique. This is done using the following steps:

–   Obtain an initial reasonable guess of the aquifer drawdown.
–   Compute the aquifer drawdown near the interface.
–   Compute the aquitard drawdown over the entire aquitard domain using relations (7.2) and (7.3).
–   Compute the aquitard vertical flux near the interface.
–   Using relations (7.1) and (7.4) compute the aquifer drawdown.
–   Repeat the iterative procedure until convergence of the respective drawdowns is achieved.

Unfortunately, when solving this class of problems with integral transforms, the solution variables can not be evaluated exactly at the interface due to Gibbs phenomenon at the end point $z = b$. In computations, the interface flux is evaluated at a small distance away from the interface. The Lanczos convergence factor discussed in Hamming [16], was found to modulate the Gibbs phenomenon in this particular problem.

8. RESULTS. The first term in the aquifer drawdown solution (7.1) is the aquifer response to pumping. The second term is the contribution of the confining layer. The aquifer may recover, be depleted (the model is rendered invalid), or reach steady state depending upon the hydrodynamic balance between the aquifer response to pumping and the response of the confining layers. The determination of such balance is an important step toward a reasonable simulation of the aquifer system drawdown. Most significantly, the balance conditions are dependent on the system hydraulic conductivities, storage coefficients, geometry, and pumping rate.

Given the aquifer parameters shown in Table 1, the coupled drawdown in the aquifer and the overlying aquitard is obtained. The aquifer material exhibits anisotropic characteristics, whereas the aquitard has no radial hydraulic conductivity components. As shown in figure 2, the exact solution to the aquifer system drawdown at R = 10 ft is significantly different from the Hantush approximate solution to the problem. The depth averaged drawdown is 1.74 ft according to the Hantush solution, whereas the exact depth averaged drawdown is 1.40 ft. For the case where the aquitard radial hydraulic conductivity is significant, the system drawdown at R = 10 ft at various aquitard hydraulic parameters is shown in figure 3. The system parameters are shown in Table 2. As seen there, only when the aquitard hydraulic parameters are near one order of magnitude less than those of the aquifer, need the leakage from the aquitard be considered. Otherwise, it is sufficient to use the Hantush impermeable aquifer solution for aquifer drawdown simulation and parameter estimation. This observation corroborates previous results reported by Neuman and Witherspoon [5].

Using the aquifer system parameters shown in Table 2, case 1, the resulting velocity field is shown in Figure 4. This evidently compares favorably with Figure 5 which shows the velocity field for the same problem using a finite element solver. In contrast to the classical concept of a strictly vertical flow in the aquitard, the figures show strong radial flow velocity components in the aquitard as well as the aquifer.

VARIABLE	AQUIFER	AQUITARD
Radial Hydraulic Conductivity [ft/d]	8.6	0.0
Vertical Hydraulic Conductivity [ft/d]	0.86	0.086
Specific Storage [1/ft]	0.001	0.0
Thickness [ft]	12.5	25.5
Discharge [ft^3/d]	305.0	
Pump Screen Length [ft]	5.0	

Table 1. Aquifer System Input Requirements for Aquifer Solution Shown in Figure 2

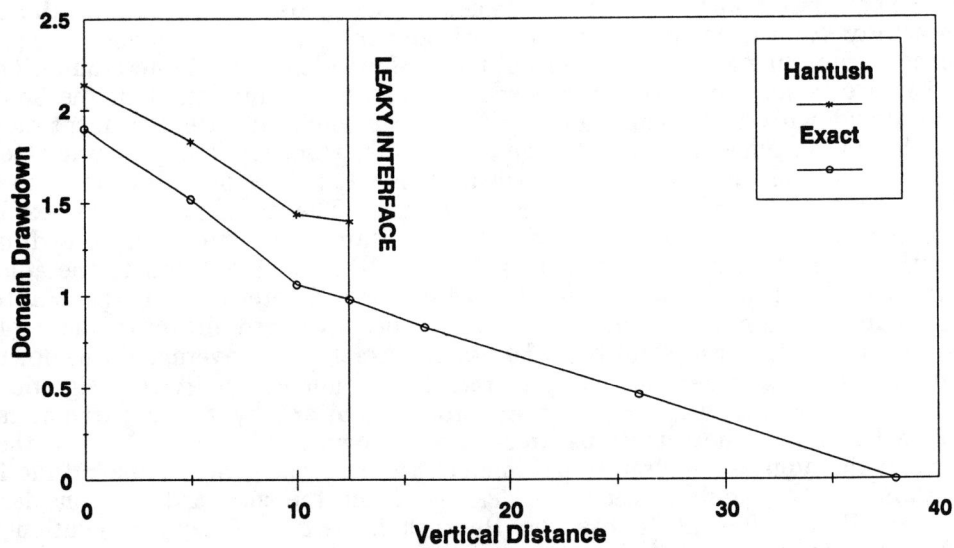

Figure 2. Aquifer Drawdown at R=10 ft, with Hydraulic Parameters as per Table 1.

VARIABLE	AQUIFER	AQUITARD CASE STUDIES		
		1.	2.	3.
Radial Hydraulic Conductivity [ft/d]	8.6	8.6	0.86	0.086
Vertical Hydraulic Conductivity [ft/d]	0.86	0.86	0.086	0.0086
Specific Storage [1/ft]	0.001	0.0	0.0	0.0
Thickness [ft]	12.5	25.5	25.5	25.5
Discharge [ft^3/d]	305.0			
Pump Screen Length [ft]	5.0			

Table 2.   Aquifer System Input Requirements to Solutions Shown in Figures 3 through 5.

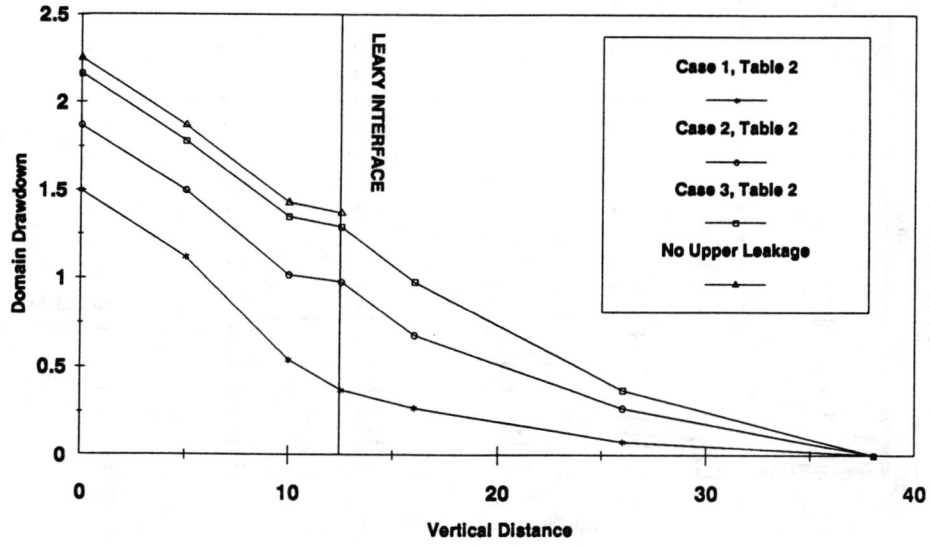

Figure 3.   Aquifer Drawdown at R=10 ft, with Aquifer Hydraulic Parameters as per Table 2, Cases 1, 2, 3.

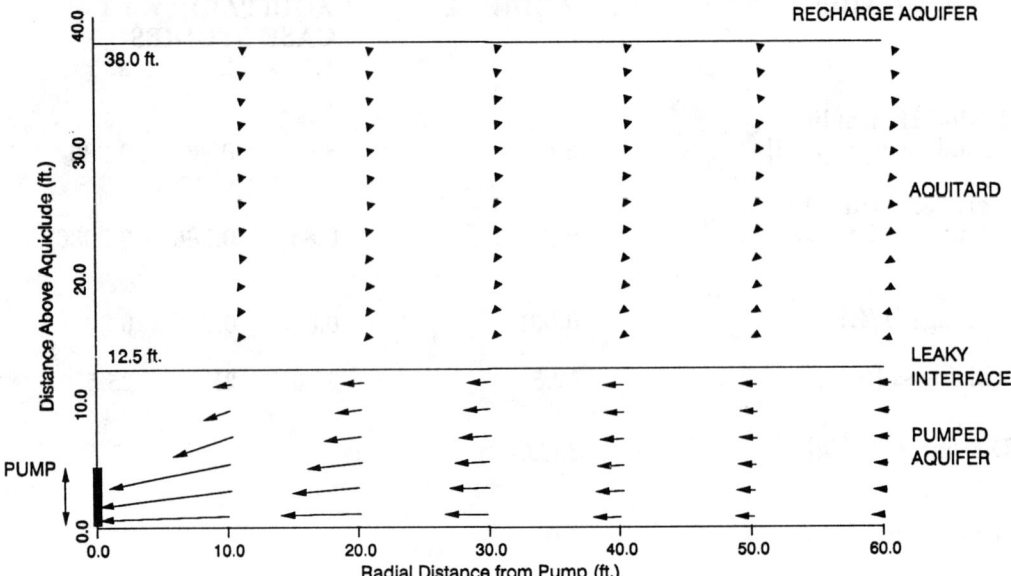

Figure 4. Velocity Vector Field Generated using the Exact Analytical Model with Input as per Table 2, Case 2.

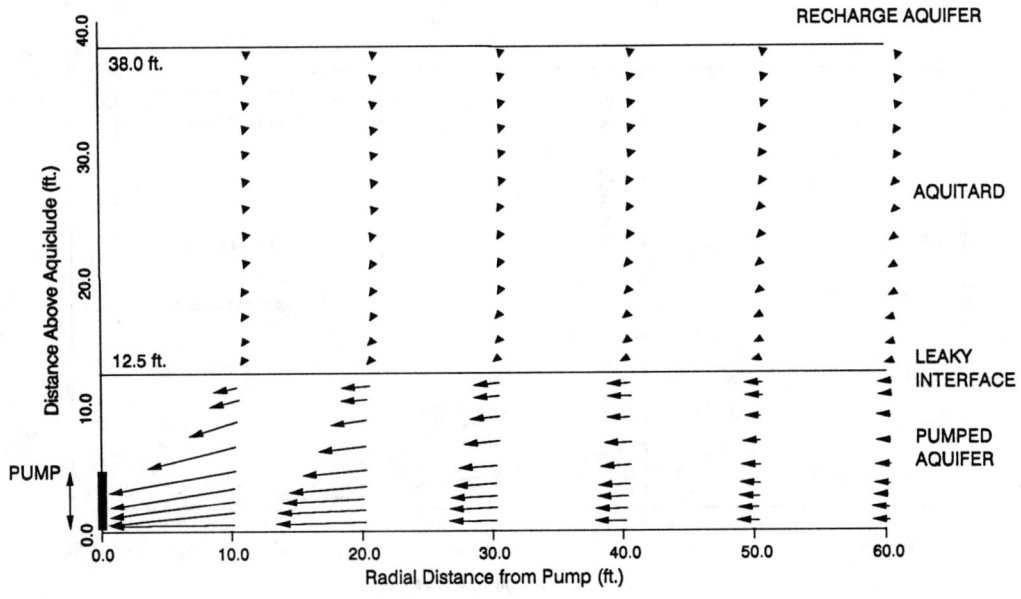

Figure 5. Velocity Vector Field Generated using a Finite Element Simulation with Input as per Table 2, Case 2.

9. DISCUSSION. An exact solution for the hydrodynamics of a leaky aquifer is obtained.   Nontrivial differences do exist between this solution and previous solutions appearing in the literature.   The solution developed here determines the system behavior and provides a complete analytical representation of the flow within the aquifer system confining layers.   This representation should provide for additional insight into the behavior of leaky aquifer systems without the need of exhaustive case studies using numerical solvers.   The analytical results are supportive of the results of Chen [6] who noted the existence of a radial component of flow in the aquitard of a leaky aquifer system.   Furthermore, the results provide for the determination of the radial flow in the aquitard explicitly.   The rigorous solution can be directly extended to multi–layered aquifer systems using the same Green's function.   Finally, although existing interpretations of the behavior of pumped leaky aquifers suggest that a steady state is ultimately achieved, the exact mathematical expression suggests that an aquifer system may or may not tend to steady state depending on the system parameters.

REFERENCES

1. C. V. Theis, *The Relation Between the Lowering of the Piezometric Surface and the Rate and Duration of Discharge of a Well using Groundwater Storage.* Trans. Amer. Geophys. Un., 16, (1935), pp. 519–524.

2. M. S. Hantush and C. Jacob, *Non–Steady Radial Flow in an Infinite Leaky Aquifer.* Trans. of the Amer. Geophys. Un., V. 36, (1955), pp. 95–112.

3. M. S. Hantush, *Hydraulics of Wells,* in the *Advances in Hydroscience,* edited by Van Te Chow. V. 1,. Academic Press, N.Y. 1964, pp. 282–430.

4. M. S. Hantush, *Analysis of Data from Pumping Tests in Anisotropic Aquifers.* J. of Geophys. Res. 71, (1966), pp. 421–426.

5. S. P. Neuman and P. A. Witherspoon, *Theory of Flow in a Confined Two Aquifer System.* Water Res. Res., V. 5, (1969), pp. 803–816.

6. Z. Chen, Z. Pang, L. Jiang, and M. Liu, *Exact Solution for the Problem of Crossflow in a Bounded Two–Aquifer system With an Aquitard..* Water Res. Res., V. 22, No. 8, (1986), pp. 1225–1236.

7. I. Herrera, *Integrodifferential Equations for systems of Leaky Aquifers and Applications.,* water Res. Res., V. 10, No.4, (1974), pp. 811–820.

8. E. O. Frind, *Exact Aquitard Response Functions for Multiple Aquifer Mechanics.* Adv. in Water Res., Volume 2, (1979), pp. 77–82.

9. R. A. Freeze and J. A. Cherry, *Ground Water,* Prentice Hall, N.J., 1979

10. M. Abramowitz, and I. Stegun, *Handbook of Mathematical Functions with Formulas, Graphs, and Mathematical Tables,* U.S. Dept. of Commerce, Applied Mathematics Series 55, U.S. Government Printing Office, Washington, D.C., 1964.

11. J. M. Bownds, *A Formal Solution to a Classical Initial–Boundary Value Problem in GroundWater Hydraulics.* Appl. Math. Comp., V. 26, (1988), pp. 333–354.

12. J. M. Bownds, *A Green's Function for a Two Point Boundary Value Problem in Groundwater Theory.* Diff. and Int. Eqs., V. 1. No. 2, (1988), pp. 173–181.

13. J. M. Bownds, *Computation of the Forcing Term for Boundary Flow through a Permeable Boundary.* J. Math. Anal. and Appl., V. 135, No.2, (1988), pp. 702–711.

14. J. M. Bownds, and T.A. Rizk, *The Inverse Laplace Transform of an Exact Analytical Solution for a BVP in Groundwater Theory.* J. Math. Anal. Appl., to appear.

15. T. A. Rizk, *Groundwater Flow and Contaminant Transport in Geologic Media,* Ph.D. Dissertation, Mech. Eng. Dept., NCSU, Raleigh, N.C., 1990.

16. R. W. Hamming, *Numerical Methods for Scientists and Engineers,* McGraw–Hill, N.Y., 1962.

CHAPTER 10

Predictive Computer Simulation of Clastic Sedimentary Processess

Daniel M. Tetzlaff*

Abstract

The mechanisms of erosion, transport, and deposition of clastic sediment produce many of the geologic structures that are of interest to petroleum geologists. These mechanisms are well understood at the scale of individual sediment grains. Large-scale sedimentary features (such as meanders, deltas, and turbidite fans) are driven by the same fluid-grain interactions, but have eluded modeling with a unified mathematical system. They have been treated either conceptually, or through empirical models that are applicable only to specific sedimentary environments.

A general model (SEDSIM) has been developed to simulate free-surface transient flow over an arbitrary topography, and sediment transport by the moving fluid. It represents velocity and depth in two horizontal dimensions, while simulating sedimentary deposits in full three dimensions. A set of partial differential equations describes the local interaction between the variables of the simulated system. Initial conditions include topography, sediment properties, and location, intensity, and sediment load of a set of flow sources. A particle-mesh method has been used to solve these equations. Work is currently under way to expand the model to represent flow in full three dimensions.

The SEDSIM model can be used as a sedimentological laboratory, without the scaling problems associated with physical experimentation. Numerical approximations and limited computer power restrict the time, spatial extent, and resolution of the simulations. Although it is completely deterministic, the model exhibits chaotic behavior (i.e. small uncertainties in the initial conditions lead to significant uncertainties in the model's subsequent evolution); prediction is therefore often limited to a statistical interpretation. The key to making the model useful for characterization of hydrocarbon bearing reservoirs is to find effective ways to constrain the output to honor observations such as core samples or geophysical information from actual sedimentary sequences.

1 Introduction

The modeling of sedimentary processes has three main practical applications: (1) to determine and understand processes that occurred in the geologic past (historical

---

*Texaco, Inc., 3901 Briarpark, Houston TX 77042

geology), (2) to predict the future evolution of a sedimentary system (environmental engineering), and (3) to determine the present configuration of sedimentary deposits (exploration geology).

The large time and spatial scales involved in geological phenomena seriously limit physical experimentation. Large-scale geologic processes are sometimes reduced to sizes manageable in a laboratory, but such experiments often lack rigor because some physical variables such as fluid viscosity, fluid density, and gravity, are difficult or impossible to change.

Mathematical models implemented on digital computers, on the other hand, are restricted by the complexity of the systems represented, and by the limited power of computers. The tremendous increase in computer power over the last few years, however, has permitted the implementation of relatively complete mathematical models of sedimentation, to the point that these models now have useful practical applications [2].

Small-scale sedimentary processes have been thoroughly understood for most of this century. For example, the fluid shear stress required to start moving a grain of sediment of a given size, shape, and density, is well known and can easily be calculated [14]. The vertical velocity profile and variation of sediment content within a river have been extensively studied and are fairly well understood. Open channel flow formulas developed by engineers [9,6] can predict flow velocity and depth of a river as a function of discharge.

Yet progress has been slow in the development of a general mathematical theory that predicts the effects of these processes when they operate over large areas and throughout extended periods of time. The phenomena of flow, erosion, transport, and deposition are relatively simple when observed locally, but their long-term interaction is difficult to predict. Part of the difficulty stems from the large number of variables and interactions between them that must be defined for a realistic model. Another set of difficulties, however, stem from the inherent unpredictability of sedimentary processes. These difficulties are not caused by limited computer power nor by numerical approximations used in the implementation. To the contrary, they become more evident in more detailed and realistic models. The purpose of this paper is to reveal this type of problems and to attempt to promote the development of new mathematical tools to handle them rigorously.

2   Examples of sedimentation models

Some of the earliest theoretical sedimentation simulation models that worked at geologic scales were developed by quantitative geomorphologists. Strahler [16] described a one-dimensional model to simulate the evolution of a graded river's profile. He observed that at any particular time, the relationship between elevation of the river bed and distance from the head is given by an exponential function. Strahler also postulated that the rate at which the stream profile is lowered at a given point is proportional to the slope at that point. By solving these two simple equations, he obtained an expression in which the elevation of the river bed is an exponential function of time and distance from the head ($y = y_0 \exp(-k_0(x + k_1 t))$), where y is elevation, y_0 is initial elevation at the head, x is distance from the head, t is time, and k_0 and k_1 are constants).

Though extremely simple, Strahler's model can be used to perform "experiments". For example, by varying the resistance to erosion of the underlying bedrock (k_1) along the length of the river, different profiles are produced. Unlike other models of its time (and some recent models as well), Strahler's model does not specify the macroscopic evolution of the system "a priori". Rather, it describes local interactions between variables, from which the overall behavior results after specifying initial and boundary conditions.

Another clastic sedimentation model was developed by Harbaugh and Bonham-Carter [4], to represent sedimentation on a continental shelf. It is based on a conceptual model by Sloss [15]. Rather than being based on a set of partial differential equations, Harbaugh and Bonham-Carter's model is stated directly in terms of a discrete system. The model assumes that the continental shelf can be represented in a vertical plane as a set of columns that extends seaward from the shore. Each column is filled with sediment up to the sea floor level. At each discrete time increment, an amount of sediment is brought into the first column from a source on shore. A fraction k is retained in it, and the rest $(1 - k)$ is carried onto the next column, where the process is repeated. The quantity $-\ln(k)$ may be likened to a sediment dispersion constant. An additional rule specifies that sedimentation cannot take place above the wave-base, which is a surface of constant elevation slightly below sea level. If the amount of sediment available for deposition exceeds the amount needed to fill to wave-base level, then only enough sediment to reach the wave-base level is deposited, and the rest is carried to the next column.

Many variations on the basic model are possible. For example, the amount of sediment at the source can be varied through time. The model can be made to handle multiple sediment types (such as sand, silt, and clay) simultaneously; each type has a specific value of k, and the assumption is made that coarse sediments are deposited before finer types. The capability of simulating crustal subsidence due to sediment load can also be incorporated. Subsidence rate may be kept constant throughout the section, or it may be varied locally, within each column, as a function of sediment load in that column. A time lag can be introduced so that subsidence occurs a certain number of time increments after a new load is received.

Harbaugh and Bonham-Carter's model produces deposits that strongly resemble those formed on actual continental shelves. Many other capabilities have been added to this model, and it has served as a precursor to more recent models currently used in the oil industry. In most of these models, however, spatial variability in the simulated deposits results only from significant changes in input conditions. For example, considerable changes in sea level or climate are necessary to explain a vertical change in the lithologic characteristics of the deposits. While sea level and climate have been shown to vary widely in the geologic past, many geologic processes, such as delta lobe advancement and abandonment, river meandering, river avulsion, and others, generate considerable variability with steady sources and boundary conditions. The SEDSIM model was developed with to help understand these processes.

3   The SEDSIM model

SEDSIM simulates free-surface flow in two horizontal dimensions, taking into account flow depth, and using vertically averaged flow parameters. The flow equations used by SEDSIM were derived from simplification and reduction to two dimensions of the Navier Stokes equations. The model operates in a rectangular area over which a topographic surface (z) is defined. Flow velocity ($\vec{q}$) and flow depth (h) are represented in two horizontal dimensions. Two equations describe fluid flow: the flow continuity equation and the flow kinematic equation. The flow continuity equation states

$$\frac{\partial h}{\partial t} = \nabla \cdot (h\vec{q}) \tag{1}$$

where

$h =$ flow depth ($h \geq 0$),

$t =$ time,

$\vec{q} =$ flow velocity vector (in two horizontal dimensions).

The flow kinematic equation states:

$$\frac{D\vec{q}}{Dt} = -g\nabla(h + z) + \frac{\mu}{\rho}\nabla^2\vec{q} - c\frac{\vec{q}|\vec{q}|}{h} \tag{2}$$

where

$\frac{D}{Dt} =$ Lagrangian time derivative, equivalent to

$$\frac{\partial}{\partial t} + (\vec{q} \cdot \nabla)$$

$g =$ gravitational acceleration,

$z =$ topographic elevation,

$\mu =$ fluid viscosity,

$\rho =$ fluid density,

$c =$ bottom-friction constant.

The initial conditions for the flow model are given by the initial topographic elevation z_0 at every point, the initial flow depth h_0, and the initial flow field $\vec{q}_0$.

Boundary conditions for the flow model are given at the "shoreline" and at the edges of the simulated area. The shoreline is the line at which depth h becomes 0; it is a free boundary that moves with the local flow velocity

$$\frac{Dh_s}{Dt} = 0 \tag{3}$$

where

h_s = flow depth at the shoreline.

At the edges of the simulated area the flow is allowed to exit freely. It is convenient to assume that the derivative of flow depth perpendicular to the boundary is 0, thus:

$$\vec{N} \cdot \nabla h_b = 0 \qquad (4)$$

where

$\vec{N}$ = unit vector normal to the boundary,

h_b = flow depth at the boundary.

Sources of fluid may be present within the simulated area. They are represented by small areas in which the flow continuity equation (1) does not hold, but is replaced by the following relationship:

$$\frac{\partial h}{\partial t} + \nabla \cdot (h\vec{q}) = F_i \qquad (5)$$

where

F_i = source intensity per unit area.

The source intensity multiplied by the source area yields the source flow (Q_{S_i}).

If the topographic surface z is constant through time, the above equations, initial conditions, and boundary conditions define a complete two-dimensional flow model. The addition of equations governing sediment erosion, transport, and deposition allows the topography to change with time and incorporates a new variable, sediment load, as a scalar property at every point in the horizontal plane. The sediment equations consist of a sediment continuity equation and a sediment kinematic equation. The sediment continuity equation states:

$$h\frac{Dl}{Dt} = -\frac{\partial z}{\partial t} \qquad (6)$$

where

l = sediment load, defined as the volume of sediment per unit volume of fluid (dimensionless),

z = topographic elevation.

The sediment kinematic equation states:

$$h\frac{Dl}{Dt} = \begin{cases} (\Lambda - l/k_1)k_2 & if\ \Lambda - l/k_1 < 0\ or\ \tau_0 > \tau_c \\ 0 & otherwise \end{cases} \qquad (7)$$

where

Λ = transport capacity, equal to $c_t g n^2 \rho |\vec{q}|^3 / h^{4/3}$, where c_t is a transport coefficient and n is Manning's roughness coefficient,

k_1, k_2 = constants that depend on sediment properties,

τ_0 = bottom shear stress, defined as $c|\vec{q}|^2/h$,

τ_c = critical shear stress (minimum shear stress needed to pick up sediment from the bed).

Equation (7) was deduced as an extension of a commonly used steady-state transport equation originally formulated by Meyer-Peter and Muller [11]. Parameters n, and τ_c are empirical and are of common use in engineering; k_1 and k_2 are also empirical and were determined for this model from published experimental and field data; c_t is chosen arbitrarily (1 ms^3/kg) since it only represents a scale factor for k_1 and k_2.

The sediment equations also require initial and boundary conditions. The initial sediment load (l_0) must be given at every point where $h > 0$. The flow sources are assumed to bring in sediment. Therefore, it is also necessary to specify the sediment load L_i at each source i.

The sediment load is assumed to leave the area at the edges with the fluid that transports it. The model assumes that the derivative of sediment load perpendicular to the boundary is 0:

$$\vec{N} \cdot \nabla l_b = 0 \tag{8}$$

To preclude excessive erosion or deposition at the edges, an additional edge boundary condition specifies that the topographic elevation is not allowed to change there. Thus:

$$\frac{\partial z_b}{\partial t} = 0 \tag{9}$$

A particle-mesh numerical method [5,7] is used to solve this system. It represents the fluid by a large number of small elementary volumes moving over a fixed grid. Though this approach is computationally lengthy, it was used because it facilitates handling highly unsteady flow. SEDSIM has also been expanded to work with several sediment types. Further details on the numerical method and the extension to multiple sediment types are provided by Tetzlaff and Harbaugh [17]. The possibility of expanding the model to simulate flow in full three dimensions is presently being studied. Through implementation on parallel computers, the task appears to be possible.

4   Use of SEDSIM

The SEDSIM computer program is lengthy to execute but extremely flexible. It can simulate steady or unsteady flow in a variety of clastic environments (continental, marine, or mixed). Sedimentary deposits are represented in full three dimensions and can be displayed graphically as maps, sections, and perspective views. Successions of displays show the evolution of the simulated systems through time. SEDSIM reproduces many sedimentary features that occur in a variety of environments and flow conditions. It has been used in a number of geological experiments [13,17] spanning up to 50,000 years simulated time.

A simple experiment is presented below. It simulated the formation of alluvial fans, and assumed an initial topographic surface consisting of a gently sloping plane broken by a steep escarpment (Fig. 1a). The extent of the simulated area was 3 km by 2 km. The sediment present in the area was poorly consolidated sand. In the first experiment, two sources (S1 and S2) were placed on the upper part of the slope. The sources differed very slightly from each other ($Q_{S_1} = 19.9m^3/s$, $Q_{S_2} = 20.0m^3/s$). They were activated intermittently for 0.01 years every year to simulate yearly floods. The sediment load at both sources was 0.01. The experiment was run for a simulated time of 100 years. This is geologically very short, but it is long enough for a significant alluvial fan system to develop.

After a simulated time of 20 years (Fig. 1b), channels have been eroded on the upper part of the area, and alluvial fans have been deposited on the lower part. After 100 years (Fig. 1c) the flow has changed course, the channels have deepened, and the deposits have grown. Figure 2 shows a section through the deposits at 100 years. Though the difference in flow between the two sources was very small, significant differences between the features formed by each source can be observed.

Other experiments have shown that when independent runs are made that differ very slightly in initial conditions, the results may differ significantly. These differences between runs that use very similar input may be large even when the input differs by the smallest value allowed by the machine's precision. Only when all parameters are identical, are the results identical. This type of behavior has been observed in many models of physical systems (and in the physical systems themselves), and is a manifestation of the property called chaos [12,10]. SEDSIM has not been rigorously proven to be chaotic, but its behavior strongly suggests it is. The difference in the end results stems from the small initial difference in flow conditions, which was amplified by turbulence, erosion, and deposition. This amplification is not due to artificial numerical instability, but is a property of the physical system represented. If many runs are made with initial conditions varied (randomly or systematically) within a very small interval the results can be interpreted as a statistical distribution. If we plot sediment thickness at a point P as a function of source flow Q_S (Fig. 3), we obtain a plot that appears to be erratic, although it shows a systematic trend. Even though SEDSIM is deterministic, its behavior is remarkably similar to that of a random model.

It is also noticeable that the topographic surface quickly becomes irregular, even though the initial surface was perfectly smooth (Fig. 1). During the first few days of simulated time, very small irregularities appear due to round off errors and due to the way the flow is simulated (i.e, using a finite number of fluid elements), and are later amplified. In a physical system, initial irregularities are also present due to imperfections in the topographic surface, due to unevenness in the shape and size of individual sediment grains, and due to any minor departure from the ideal case that one attempts to simulate. These irregularities are amplified by the the system to produce turns in the river, asymmetry in the deposits, and other features. The source of the initial "perturbations", however, is very different in a physical system than in a computer-simulated system. SEDSIM's results appear to be statistically similar to those of physical systems, but there is no rigorous proof of whether the numerical errors introduce a systematic bias in the results.

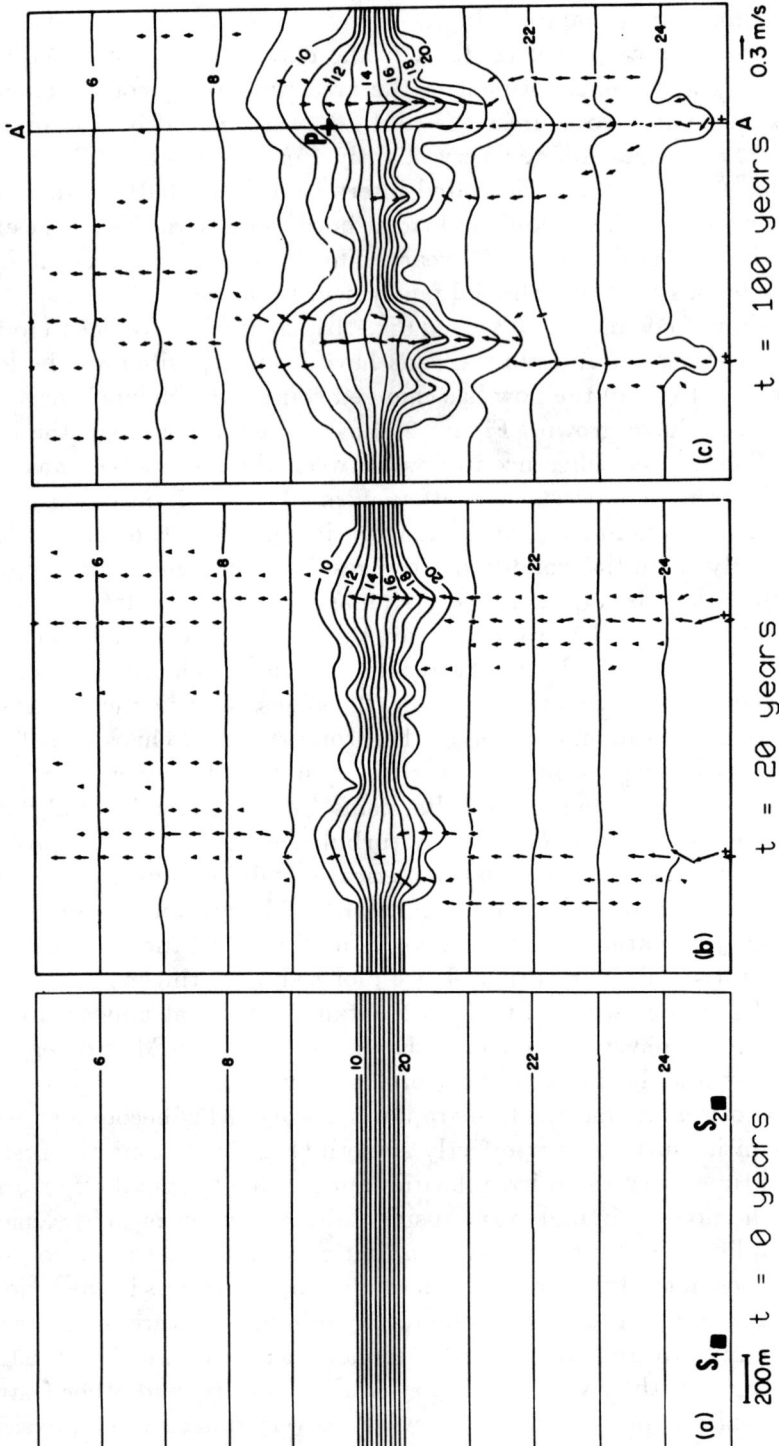

Figure 1: Plan view of sedimentary deposits in simulation experiment (contours in meters); (a) at time 0, (b) after 20 years, (c) after 100 years.

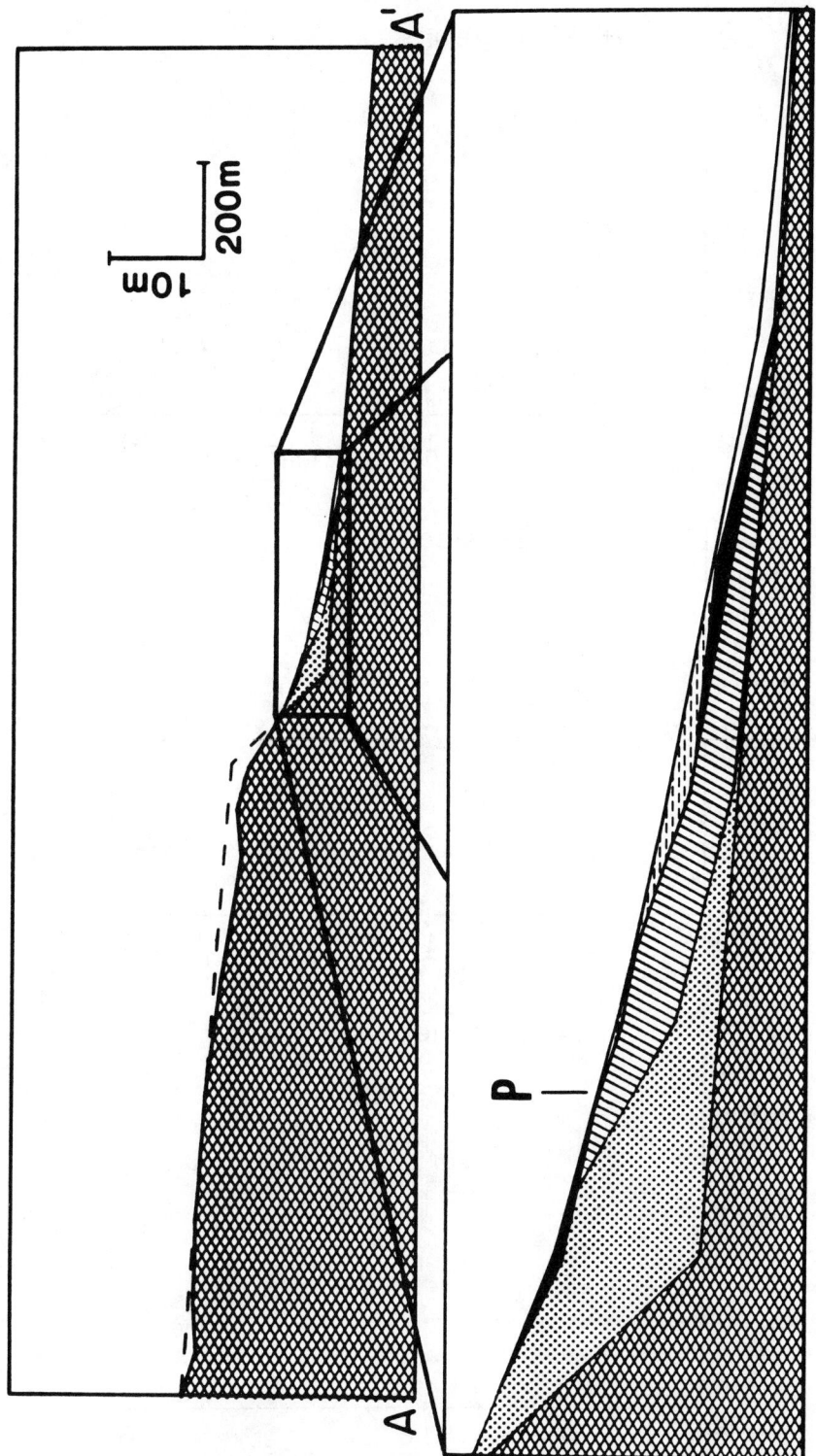

Figure 2: Section through simulated deposits at 100 years, and location of measurement point P.

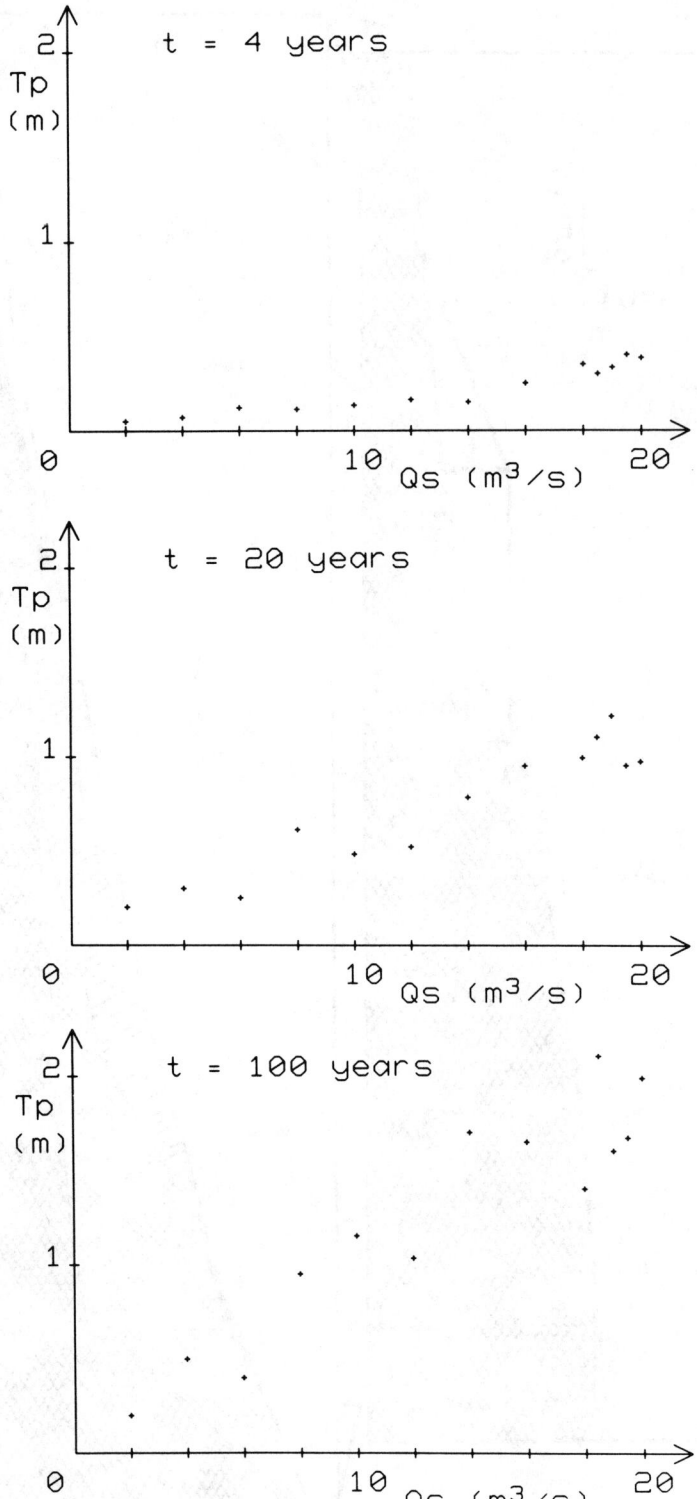

Figure 3: Plots of source flow (Q_S) vs. sediment thickness at P (T_P) for different times.

The amplification of numerical perturbations causes SEDSIM to act as a pseudo random number generator, the initial parameters being the "seeds". The "shuffling" caused by chaotic behavior causes the widely differing results between runs that employ initial parameters that differ only minimally. In this sense, SEDSIM can be considered a random model. Its behavior blurs the distinction between deterministic and random computer models.

In order to simulate the geologic past, one might ask whether SEDSIM can be run backwards from data about the present. It is well known that many models of physical systems cannot be run backward in time because information about the past is lost as the system evolves. For example, given the position and velocity of a ball on a pool table, one can trace its movement backward for a few seconds. But friction has a significant effect, and information about the ball's past history is progressively lost. We must know the ball's position and velocity with greater accuracy if we are to retrace its movement backward, and since that knowledge cannot be perfect, we can only determine its past history in a probabilistic sense. Similarly, geologic features do not retain all the information on how they were formed, and in general, SEDSIM cannot be run backwards as a deterministic model. But furthermore, SEDSIM cannot rigorously be run forward (in a deterministic sense) either. We never have enough information about the present state of a real system to be able to predict its behavior unequivocally for a long period of time. We can only treat the problem with a probabilistic approach. For example, in the alluvial-fan experiment, we can still exploit the trend of the plot to obtain meaningful predictions, although these predictions will be probabilistic rather than exact.

The main problem for practical application is to obtain statistical predictions that are constrained or conditioned by observations. An exploration geologist, for example, might want to estimate the thickness of a formation at a particular location, given a few thickness measurements nearby, and given an approximate knowledge of the processes that formed the deposit. Three groups of techniques that attempt to achieve constrained predictions are described below. They are:

1. Pseudoinverse modeling

2. Statistical modeling

3. Spatial correlation

5   Pseudoinverse modeling

One possible way to overcome the problem of incomplete knowledge of initial conditions in conventional (non-chaotic) forward models is to use "pseudoinverse" modeling. Pseudoinverse modeling is a trial-and-error procedure in which a model's input is progressively adjusted in successive computer experiments in an attempt to achieve a closer match between simulated and actual geologic features.

Almost all of the practical applications of SEDSIM, and of other simulation models, are presently done using this technique. The method helps to understand geologic processes and predict properties of sedimentary deposits, but it is highly intuitive and does not provide reliable quantitative predictions nor error estimates.

Pseudoinverse modeling can be tedious and expensive when each experiment involves extensive computations. It is also partly subjective because intuition is often used in deciding how to adjust the input to improve the results. To reduce the subjectivity in pseudoinverse modeling, it is possible to make it automatic by employing a minimization procedure. First, we need to define a function that represents, as a single number, the error between the simulation's results and the actual features. Then we need to find the combination of input parameters that minimizes the error function. The search can be done using minimization procedures that do not require knowledge of the derivatives of the error function [1]. The error function may be difficult to define, and depends on which characteristics of the simulated features are considered to be most important.

6   Statistical modeling

Statistical modeling is somewhat more rigorous than pseudoinverse modeling. Suppose we are interested in the value of a variable T that takes different values within the simulated system. The variable T could represent, for example, thickness of sediment accumulated over a period of 100 years. Suppose the value of T is known at point P and is desired at point X. For simplicity, assume also that we know all conditions except source intensity Q_S. We could then perform a set of computer experiments in which only the source intensity is varied (Fig. 3), and thus obtain a function that assigns a value of T_P to every possible source intensity. The set of experiments will also yield a value of T_X for every source intensity (which we could plot in a three dimensional expansion of Figure 3, in which the third dimension would be T_X. It is then possible to estimate the conditional distribution $f(T_P, T_X | Q_S)$.

The following distributions, which can be obtained from the conditional distribution $f(T_P, T_X | Q_S)$, are useful in the three main areas of application of simulation models mentioned in the Introduction, namely:

1. $f(Q_S | T_P)$ allows to estimate the probability of a range of source intensities given the observation at P. This type of distribution would be of use to the historical geologist in that it permits to estimate a the past (source intensity Q_S) from an observation in the present. To estimate this distribution, it is necessary to assume the a priori distribution of Q_S, but in general, the uncertainty in Q_S given T_P (i.e. standard deviation of $f(Q_S | T_P)$) is not sensitive to a large uncertainty in the a priori distribution.

2. $f((T_P, T_X) | Q_S)$ allows to estimate property T at two points given the source intensity. This distribution is of use to the engineer, in that it allows to predict properties of future sedimentary deposits from knowledge of present conditions.

3. $f(T_X | T_P)$ allows to estimate property T at point X given an observation at point P. This estimation is of interest to the exploration geologist, who is con-

cerned with the present properties of a sedimentary deposit, given a reduced set of observations. The exploration geologist is ultimately not concerned with the source intensity S, except as a means of enabling him to predict sediment properties at present, but as in case 1., an assumption about the a priori distribution of Q_S is necessary.

A major difficulty of this approach is that in practice many more than three variables come into play. The statistical distributions described are an oversimplification. T_P must be replaced by a large set of variables whose values have been measured in an actual system, T_X would represent a set of unknown variables, and Q_S would represent a set of variables that can be controlled, and modified for each simulation experiment. When many variables are included in the system, the number of experiments needed for a reliable estimate of the joint statistical distribution increases very rapidly, to the point of becoming prohibitive from a computational standpoint.

Both pseudoinverse modeling and statistical modeling benefit from a reduction in the number of variables used. The challenge is to define a reduced set of variables that are still meaningful for the purpose of the study. For example, instead of being concerned with sediment thickness at a single point in the system, one might use average thickness over an area. Average thickness shows less variability as a function of model input parameters, thereby reducing the number of necessary experiments, but it also reduces resolution.

When using models that involve more than one sediment type, we are often concerned with the vertical distribution of a sediment property. For example the exploration geologist may want to know whether the shale content increases or decreases upward in a complex sand-shale sequence, without being concerned about individual beds. In this case, it is useful to employ the vertical moments of a sediment property, defined as follows:

$$M_i(P) = \int_{-1}^{+1} z^i R(z) dz \tag{10}$$

where

$M_i(P)$ = moment of order i for property R at point P,

z = depth scaled so that -1 represents the bottom of the sequence and +1 represents the top,

$R(z)$ = observed property at depth z for point P,

P = point in the horizontal plane.

For pseudoinverse modeling, a useful error function that can be defined for sand shale sequences using the concept of vertical moments is the following:

$$E = \int_A [M_i(P) - m_i(P)]^2 dP \tag{11}$$

where

A = area of simulated system,

n = maximum order of moments used,

$m_i(P)$ = moment of order i for simulated system at point P (defined like M_i, but for simulated rather than observed property).

The value of n can be chosen at will. A value of 1 will lead to agreement in average properties only; higher values of n will take into account vertical variations in the property with progressively greater detail.

7   Spatial correlation

Statistical techniques that employ a spatial correlation assumptions to interpolate physical properties are well known in exploration geology. For example, the techniques used in geostatistics [8] are based on the assumption of existence of a variogram function, which provides a measure of the correlation between two points as a function of distance between them. This function is generally assumed to be steady throughout the volume of rock or sediment being studied, but is often hard to establish due to lack of observations.

A simulated deposit provides complete three-dimensional information about sediment properties, permitting the determination of a spatial correlation function. If the geologist approximately knows the sedimentary environment and conditions in the area of study, then the model can be run several times with a set of conditions that approximately covers the range of values assumed to have existed in the geologic past. The results may be analyzed to estimate spatial correlation function for the variables of interest.

After a spatial correlation function has been obtained, it can be used to perform random simulations in Monte Carlo fashion, to produce a number of equiprobable outcomes. Each outcome is slightly different, yet all possess the same spatial variability function, and all strictly honor observations. These techniques of conditional simulation of spatial variability are computationally much faster than sedimentary process simulation. The main disadvantage variogram-based conditional simulation that a variogram often does not contain enough information to characterize a complex sedimentary deposit. Correlation functions that are more useful than the variogram used in conventional geostatistics have been investigated [3]. They may be able to reproduce detailed features of the sedimentary record, still strictly honoring all observations.

8   The need for new mathematical tools in geology

The exploration geologist generally wants to determine the present-day distribution of physical properties of sedimentary rocks with the greatest possible accuracy and spatial resolution, from a very limited set of observations. Knowledge of the geologic past is one more way to help achieve this goal. Given the necessity of a statistical approach imposed by the chaotic nature of clastic sedimentary systems, the best scenario the explorationist could hope for can be outlined as follows:

1. Obtain a statistical distribution of the past geological geological conditions that could have produced the observed data.

2. Use this distribution as input to a statistical forward model, to obtain a statistical distribution, constrained by the observations, of present day properties of the sedimentary deposit.

This procedure parallels quantitatively the train of though that exploration geologists apply intuitively. In order to fully quantify the method, further research is needed to answer the following questions:

- What sedimentary systems are really chaotic, and what tests can be applied to determine their nature?

- What numerical methods are most effective to solve chaotic systems? Is it valid to replace the perturbations that occur in a physical system by perturbations that result from rounding off and discretization?

- Can statistical distributions of results of mathematical models of sedimentation be generated by methods that are more efficient than Monte Carlo simulation?

- Can spatial variability be characterized by a method that preserves enough information for use as a faster alternative to forward modeling in simulating deposits that honor observations?

- Given partial information on the present state of a system, is it possible to obtain a statistical distribution of the past conditions that may have produced the observed state?

While theoretical research in chaotic systems has seen great progress in recent years, its application to natural sedimentary systems still presents a ample opportunities for original research, much of it of direct interest to petroleum exploration and reservoir characterization.

9   Conclusions

Most limitations that affect a model's realism ultimately reside in two causes: (1) the computer power available to run the model, and (2) the problems associated to chaotic behavior. The first is the usual limiting factor at present. The second will become more important as models become more realistic.

"Randomness", in a broad sense, can appear even in models that are defined to be completely deterministic. Processes such as fluid turbulence are deterministic because they are governed by precise physical laws, but their behavior can be predicted only within certain limits. These limits cannot be reduced by increasing the precision with which we know the initial conditions and boundary conditions.

Such processes may pose an absolute limit to the predictive ability of dynamic models.

The entire field of sedimentation simulation could benefit from the development of new mathematical tools. These tools are particularly needed in defining the relationships between deterministic chaotic models and random models.

Acknowledgments

I thank Texaco, Inc. for encouraging the presentation of this paper. I also thank Prof. John W. Harbaugh of Stanford University, who directed the development of computer program SEDSIM, and Dr. Mary Wheeler of the University of Houston and Dr. Martin Perlmutter of Texaco for their constant interest in mathematical models of geologic processes.

References

[1] R. BRENT, *Algorithms for Minimization Without Derivatives*, Prentice Hall, Englewood Cliffs, New Jersey, 1973.

[2] T. A. CROSS, ed., *Quantitative Dynamic Stratigraphy*, Prentice Hall, Englewood Cliffs, New Jersey, 1989.

[3] P. M. DOYEN, T. M. GUIDISH, AND M. H. DE BUYL, *Seismic discrimination of lithology in sand/shale reservoirs, a Bayesian approach*, in Proceedings of the 1989 Convention of the Society of Exploration Geophysicists, 1989.

[4] J. W. HARBAUGH AND G. BONHAM-CARTER, *Computer Simulation in Geology*, Krieger Publishing Company, Malabar, Florida, 1970. reprint edition 1981.

[5] F. HARLOW, *The Particle-in-cell Method for Fluid Dynamics*, vol. 3, Academic Press, New York, 1964, pp. 319–343.

[6] F. HENDERSON, *Open Channel Flow*, Macmillan Pub. Co., New York, 1966.

[7] R. HOCKNEY AND J. EASTWOOD, *Computer Simulation Using Particles*, McGraw-Hill, New York, 1981.

[8] A. J. JOURNEL AND C. HUIJBREGTS, *Mining Geostatistics*, Academic Press, New York, 1978.

[9] R. MANNING, *Flow of water in open channels and pipes*, in Trans. Inst. Civil Engrs. (Ireland), vol. 20., Ireland, 1890.

[10] G. MAYER-KRESS, *Dimensions and Entropies in Chaotic Systems*, Springer Verlag, Berlin, 1986.

[11] E. MEYER-PETER AND R. MULLER, *Formulas for bed load transport*, in Proceedings, Third meeting of Intern. Assoc. Hydr. Res., Stockholm, 1948, pp. 39–64.

[12] R. SCHUSTER, *Deterministic Chaos, an Introduction*, VCH Publishers, Inc., 1975.

[13] N. SCOTT, *Modern vs. ancient braided stream deposits: A comparison between simulated sedimentary deposits and the ivishak formation of the prudhoe bay field, alaska*, Master's thesis, Department of Applied Earth Sciences, Stanford University, 1986.

[14] I. SHIELDS, *Anwendung der Ähnlichkeitsmechanik und der Turbulenzforschung auf Geschiebebewegung*, in Mitteilungen der Preussischen Versuchsanstalt für Wasserbau und Schiffbau, Heft 26, Berlin, 1936. Available also in a translation by W.P. Ott and J.C. van Uchelen, S.C.S. Cooperative Laboratory, California Institute of Technology, Pasadena, Calif.

[15] L. SLOSS, *Stratigraphic models in exploration*, Journal of sedimentary petrology, 32 (1962), pp. 415–422.

[16] A. STRAHLER, *Mathematical models in geomorphology*, Geol. Soc. of Am. Bulletin, 63 (1952), pp. 923–938.

[17] D. M. TETZLAFF AND J. W. HARBAUGH, *Simulating Clastic Sedimentation*, Van Nostrand-Reinhold, series in Mathematical Geology, New York, 1989.

CHAPTER 11

Integral Method Solutions to Flow Problems in Unsaturated Porous Media*

Robert W. Zimmerman**
Gudmundur S. Bodvarsson**

Abstract. The integral method is used to derive approximate solutions for the problem of absorption of water into an initially unsaturated porous medium. This problem is governed by a nonlinear diffusion equation, for which exact solutions are generally not obtainable. Approximate solutions are obtained for media with two different commonly-used sets of characteristic curves, those of Brooks-Corey and van Genuchten-Mualem. The approximate solutions compare reasonably well with numerical results, and also have the advantage of clearly displaying the manner in which the solutions depend on the hydrological parameters of the problem.

Introduction. The integral method for deriving approximate solutions to nonlinear partial differential equations that arise in engineering and the physical sciences was introduced by Pohlhausen [1] to treat the problem of laminar flow over a flat plate. Pohlhausen approximated the velocity distribution through the boundary layer by a low-order polynomial whose coefficients depended on the unknown boundary-layer thickness. Although the polynomial did not satisfy the governing momentum equation exactly, it was forced to satisfy the integral of this equation over the boundary layer thickness. This led to a simple ordinary differential equation that governed the thickness of the boundary layer along the length of the plate. The approximate solution thus derived, using only a quartic profile, compared reasonably well with the exact numerical solution of Blasius [2]. In particular, the approximate solution predicted certain properties of interest, such as the skin friction, to within 3% of the exact value.

*This work was done under U.S. Department of Energy Contract No. DE-AC03-76SF00098, administered by the Nevada Operations Office, in cooperation with the U.S. Geological Survey, Denver.
**Earth Sciences Division, Lawrence Berkeley Laboratory, University of California, Berkeley, CA 94720.

The integral method seems to have been first brought to bear on diffusion problems by Landahl [3], who in the succeeding years derived approximate solutions to many diffusion problems arising in biophysics. The method has since been widely used in heat conduction problems (see [4], and references therein). Since linear diffusion equations can usually be solved by classical methods such as separation of variables or Green's functions, the integral method is most useful in deriving approximate solutions to nonlinear problems for which closed-form solutions are not obtainable. One problem in the earth sciences which leads to a highly nonlinear diffusion equation is that of fluid flow in partially saturated (also called "unsaturated") porous media. In such problems, the nonlinearities are usually fairly strong, thus limiting the usefulness of perturbation methods. Of course, numerical solutions can always be obtained for such problems [5]. Numerical methods, however, have the disadvantage of not clearly showing the manner in which the solution depends on the various parameters of the problem. The integral method leads to closed-form (albeit approximate) solutions which do give insight into the effect of the various boundary conditions and constitutive parameters of the problem. The purpose of this paper is to illustrate how the integral approach can lead to simple, but relatively accurate, solutions to otherwise intractable unsaturated flow problems.

Formulation of the Problem. Horizontal flow of water in an unsaturated medium is usually thought to be governed by Richards' partial differential equation [6]:

$$\frac{\partial}{\partial x}\left[\frac{kk_r(\psi)}{\mu\phi}\frac{\partial\psi}{\partial x}\right] = \frac{\partial S}{\partial t} . \tag{1}$$

In this equation, S is the liquid saturation, which is equal to the fraction of the pore space that is filled with water. ψ is the potential, or capillary pressure, and is related to the saturation through a capillary pressure function, $S = S(\psi)$. In regions of partial liquid saturation, ψ will be negative. k is the absolute (i.e., fully-saturated) permeability of the medium, ϕ is its porosity (assumed constant), and μ is the viscosity of water. k_r is the dimensionless relative permeability function, which measures the decrease in the permeability of the medium to the presence of air in some of the pores. If hysteretic effects are neglected, k_r and S are single-valued functions of ψ. Equation (1) can also be used for the initial stages of vertical infiltration, when gravitational forces are still negligible.

Equation (1) embodies the principal of conservation of mass for the water, along with Darcy's law to relate the volumetric flux to the potential gradient. Although the physical problem of flow in an unsaturated medium actually involves both the water and the air phases, it is conventional to ignore the air by implicitly assuming it to be infinitely mobile, and at a fixed pressure of one atmosphere. Since the relative permeability and capillary pressure functions are always strongly varying functions of S, Equation (1) is highly nonlinear. Note that Equation (1) can be put into the form of a standard nonlinear diffusion equation as follows:

$$\frac{\partial}{\partial x}\left[\frac{kk_r(S)}{\mu\phi}\frac{d\psi}{dS}\frac{\partial S}{\partial x}\right] = \frac{\partial}{\partial x}\left[D(S)\frac{\partial S}{\partial x}\right] = \frac{\partial S}{\partial t} , \tag{2}$$

where $D(S) = kk_r(S)\psi'(S)/\mu\phi$. Although this form of the governing equation is frequently used in soil physics (e.g., [7]), we will find it convenient to use the form given in Equation (1).

A basic problem in the field of unsaturated flow is that of absorption from a saturated boundary at some fixed potential ψ_w into a half-space that is initially at some uniform saturation ψ_i. Without loss of generality, we can assume that $\psi_w = 0$, since the solution for $\psi_w > 0$ is related in a simple way [8] to the solution for the case $\psi_w = 0$. The boundary and initial conditions for this problem are

$$\psi(0, t) = 0 , \tag{3}$$

$$\psi(x, 0) = \psi_i , \tag{4}$$

$$\lim_{x \to \infty} \psi(x, t) = \psi_i . \tag{5}$$

The last condition reflects the fact that, at any finite time, the wetting front moving in from the saturated boundary cannot have penetrated infinitely far into the medium. Equations (1,3-5), along with expressions for the saturation and the relative permeability as functions of ψ, completely specify the problem of horizontal one-dimensional absorption.

The capillary pressure and relative permeability functions depend on the pore geometry of the medium (see [9]), and have different forms for different media. Two of the more widely-used forms for these "characteristic equations" are those of Brooks and Corey [10], and van Genuchten-Mualem [11,12]. Neither of these sets of functions are analytic, since they typically involve fractional powers of the saturation. Furthermore, the Brooks-Corey capillary pressure function is given by different algebraic expressions in different capillary pressure regimes. While these peculiarities hinder attempts to analytical solutions, that they pose no particular difficulty for the integral method.

Solution for Brooks-Corey Media. The Brooks-Corey characteristic functions are

$$\hat{S}(\psi) = \frac{S(\psi) - S_r}{S_s - S_r} = 1 \quad \text{if} \quad |\alpha\psi| \le 1 ,$$

$$= |\alpha\psi|^{-n} \quad \text{if} \quad |\alpha\psi| > 1 ; \tag{6}$$

$$k_r(\psi) = 1 \quad \text{if} \quad |\alpha\psi| \le 1 ,$$

$$= |\alpha\psi|^{-(3n+2)} \quad \text{if} \quad |\alpha\psi| > 1 , \tag{7}$$

where S_r is the residual saturation, S_s is the saturation at zero potential, $\hat{S}$ is the normalized saturation, and α is a scaling parameter that is inversely proportional to the mean pore diameter. The parameter n must satisfy the inequality $n \geq 1$, but is not necessarily an integer [13]. The normalized saturation equals 1 for all $\psi > -1/\alpha$, after which it drops off to zero as ψ decreases, according to a power law. Since no air can enter the medium unless $|\alpha\psi| > 1$, $1/\alpha$ is often called the "air-entry pressure". The relative permeability monotonically decreases from 1 to 0 as the normalized saturation decreases from 1 to 0.

Before attempting to solve this problem, it is convenient to normalize all the variables, and transform the governing partial differential equation into an ordinary differential equation by applying a Boltzmann-type transformation [14]. If we define a normalized potential as $\hat{\psi} = \alpha\psi$, and a similarity variable η as

$$\eta = \left[\frac{\alpha\mu\phi(S_s - S_r)x^2}{kt} \right]^{1/2} , \tag{8}$$

then Equation (1) is transformed into

$$\frac{d}{d\eta}\left[k_r(\hat{S})\frac{d\hat{\psi}}{d\eta} \right] + \frac{\eta}{2}\frac{d\hat{S}}{d\eta} = 0 , \tag{9}$$

and the three boundary/initial conditions (3-5) are transformed into the two conditions

$$\hat{\psi}(0) = 0 , \tag{10}$$

$$\lim_{\eta \to \infty} \hat{\psi}(\eta) = \hat{\psi}_i . \tag{11}$$

The above transformation has the effect of reducing the problem to a two-point ODE boundary-value problem, given by Equations (9-11).

The basic idea behind the integral method is to approximate the solution with some simple function that contains an adjustable parameter, and then fix the value of this parameter by requiring the solution to satisfy the differential equation in an integrated sense. An important fact about the use of the integral method is that reasonable forms for the solution can often be obtained merely by consideration of the boundary conditions and various simple properties of the governing equation. For example, note that since $\psi = 0$ at $\eta = 0$, by continuity there will be a region near the boundary where $\psi > -1/\alpha$. Equation (6) then shows that $\hat{S} = 1$ in this region, which implies that the term $d\hat{S}/d\eta$ in Equation (9) will be zero. Equation (9) then implies that $d\hat{\psi}/d\eta$ is a constant, and so ψ will drop off linearly from 0 to $-1/\alpha$. The value of η at which $\hat{\psi}$ reaches $-1/\alpha$ will be denoted by λ. The capillary pressure will continue to decrease as η increases, reaching its initial value $\hat{\psi}_i$ at some point $\eta = \lambda + \delta$. These considerations suggest the following trial profile (Fig. 1):

$$0 < \eta < \lambda: \quad \hat{\psi} = -\eta/\lambda , \quad \hat{S} = 1;$$

$$\lambda < \eta < \lambda + \delta: \quad \hat{S} = 1 - (1 - \hat{S}_i)\frac{\eta - \lambda}{\delta} , \quad \hat{\psi} = -\hat{S}^{-1/n};$$

$$\lambda + \delta < \eta < \infty: \qquad \hat{S} = \hat{S}_i \ , \quad \hat{\psi} = -\hat{S}_i^{-1/n} . \tag{12}$$

The profiles chosen for the first and third regions follow from the considerations discussed above, while a linear saturation profile is chosen in the second region merely for its simplicity.

A relationship between the parameters λ and δ can be found by requiring continuity of the capillary pressure gradient at $\eta = \lambda$. We first calculate the capillary pressure gradient in the two regions as follows:

$$\left. \frac{\partial \hat{\psi}}{\partial \eta} \right|_{\lambda^-} = \frac{1}{\lambda} , \tag{13a}$$

$$\left. \frac{\partial \hat{\psi}}{\partial \eta} \right|_{\lambda^+} = \left. \frac{d\hat{\psi}}{d\hat{S}} \right|_{\hat{S}=1} \left. \frac{\partial \hat{S}}{\partial \eta} \right|_{\eta=\lambda}$$

$$= \left[\frac{-1}{n} S^{-(1+1/n)} \right]_{\hat{S}=1} \left[\frac{-(1-\hat{S}_i)}{\delta} \right]_{\eta=\lambda} = \frac{(1-\hat{S}_i)}{n \delta} . \tag{13b}$$

Equating these two gradients yields

$$\lambda = \frac{n \delta}{(1 - \hat{S}_i)} . \tag{13c}$$

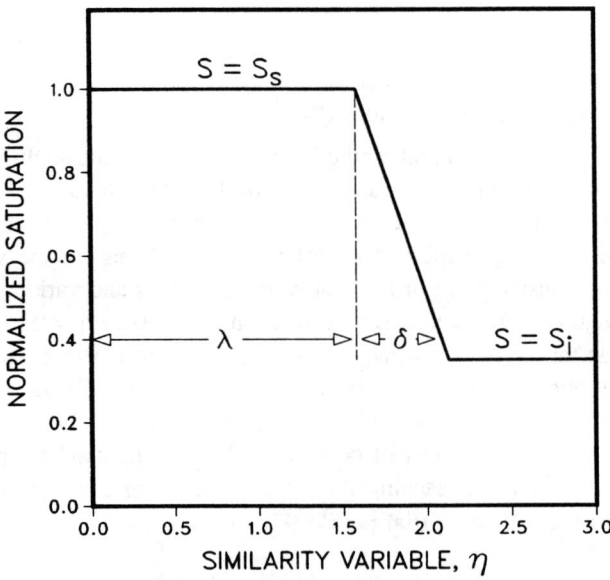

Figure 1. Assumed saturation profile for absorption into a Brooks-Corey medium, as given by Equation (12).

Note that we do not require the capillary pressure gradient to be continuous at $\eta = \lambda + \delta$; imposition of this condition leads to only slight increases in accuracy, at the expense of much additional algebraic complexity. If desired, one can imagine that there is a small "tail" on the saturation profile at $\eta = \lambda + \delta$ that smoothly connects the two piecewise-linear parts of the profile (see Fig. 1), but which is sufficiently localized so as to have no appreciable effect on the required integrals (see Equation (15)).

To find an expression for δ, we integrate Equation (9) from $\eta = 0$ to $\eta = \infty$. The first term in Equation (9) integrates out to

$$k_r(\hat{S}) \frac{d\hat{\psi}}{d\eta} \Bigg]_0^\infty = \frac{-1}{\lambda} = \frac{-(1-\hat{S}_i)}{n\delta} . \tag{14}$$

Since the term $d\hat{S}/d\eta$ is non-zero only in the range $\lambda < \eta < \lambda + \delta$, the second term integrates out to

$$\int_0^\infty \frac{\eta}{2} \frac{d\hat{S}}{d\eta} d\eta = -\int_\lambda^{\lambda+\delta} \frac{\eta}{2} \frac{(1-\hat{S}_i)}{\delta} d\eta$$

$$= \frac{-(1-\hat{S}_i)}{2\delta} \frac{\eta^2}{2} \Bigg]_\lambda^{\lambda+\delta} = \frac{-(1-\hat{S}_i)}{4\delta} [(\lambda+\delta)^2 - \lambda^2] . \tag{15}$$

Combining Equations (14) and (15) gives

$$\frac{4}{n} = \delta^2 + 2\lambda\delta . \tag{16}$$

Using Equation (13c) to eliminate λ from Equation (16) leads to

$$\frac{4}{n} = \delta^2 \left[1 + \frac{2n}{(1-\hat{S}_i)} \right] , \tag{17}$$

which can be solved for

$$\delta = 2 \left[n \left[1 + \frac{2n}{(1-\hat{S}_i)} \right] \right]^{-1/2} . \tag{18}$$

Equations (12,13,18) specify the approximate solution to the problem.

The instantaneous liquid flux into the medium, per unit surface area, can be found from Darcy's law as follows:

$$q = \frac{-kk_r}{\mu} \frac{\partial\psi}{\partial x}\bigg|_{x=0} = \frac{-k}{\mu} \frac{d\psi}{d\hat{\psi}} \frac{\partial\hat{\psi}}{\partial\eta}\bigg|_{\eta=0} \frac{\partial\eta}{\partial x}\bigg|_{x=0}$$

$$= \frac{-k}{\mu} \frac{1}{\alpha} \frac{-1}{\lambda} \left[\frac{\alpha\mu\phi(S_s - S_r)}{kt}\right]^{1/2}$$

$$= \frac{k}{\mu} \frac{(1 - \hat{S}_i)}{\alpha n \delta} \left[\frac{\alpha\mu\phi(S_s - S_r)}{kt}\right]^{1/2}$$

$$= \left[\frac{k\phi(S_s - S_i)}{2\alpha\mu t}\left[1 + \frac{(S_s - S_i)}{2n(S_s - S_r)}\right]\right]^{1/2} . \tag{19}$$

If we write the instantaneous flux $q(t)$ as $\mathbf{S}/2\sqrt{t}$, then the cumulative flux up to some time t is equal to

$$Q(t) = \int_0^t q(\tau)\,d\tau = \int_0^t \frac{\mathbf{S}}{2}\tau^{-1/2}\,d\tau = \mathbf{S}t^{1/2} . \tag{20}$$

The constant $\mathbf{S}$ is often referred to as the sorptivity [7,15]. From Equation (19) and (20), $\mathbf{S}$ can be expressed as

$$\mathbf{S} = \left[\frac{2k\phi(S_s - S_i)}{\alpha\mu}\left[1 + \frac{(S_s - S_i)}{2n(S_s - S_r)}\right]\right]^{1/2} . \tag{21}$$

In order to judge the accuracy of the approximate solution, we can compare it to the results of numerical solutions to Equations (9-11). These equations represent a two-point (ordinary differential equation) boundary-value problem, which can be solved using the shooting method [16]. Figs. 2 and 3 compare the capillary pressure and saturation profiles of the approximate solution and the numerical solution for a Brooks-Corey medium with $n = 2$, for two different initial saturations. This value of n is close to the values that have been estimated [17] for the welded tuffs at Yucca Mountain, Nevada, a potential site of an underground nuclear waste repository. The approximate solution very accurately predicts λ, the length of the fully saturated zone, but slightly overpredicts δ, the width of the partially-saturated zone. Since the sorptivity is proportional to the slope of the capillary pressure profile at $\eta = 0$, and to the area bounded by the saturation profile and the line $S = S_i$, it is clear that the approximate solution estimates $\mathbf{S}$ very accurately. Fig. 4 shows the normalized sorptivity $\mathbf{S}/[k\phi/\alpha\mu]^{1/2}$ plotted against the initial saturation, for a few different values of n. For simplicity, S_r is taken to be 0, and S_s is taken to be 1. The approximate solution is seen to estimate the sorptivity very accurately, over the entire range of initial saturations. Although the accuracy is higher for larger values of n, and for higher values of the initial saturation, in all cases the approximate sorptivity is correct to within 10%.

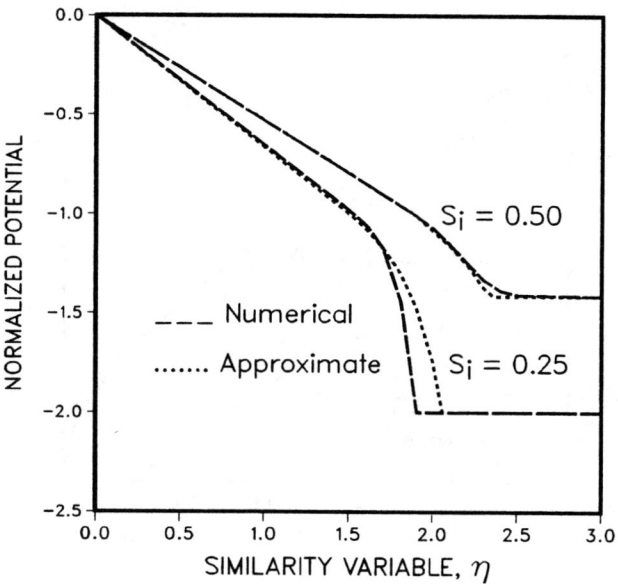

Figure 2. Potential profiles for absorption into a Brooks-Corey medium, according to the approximate and numerical solutions.

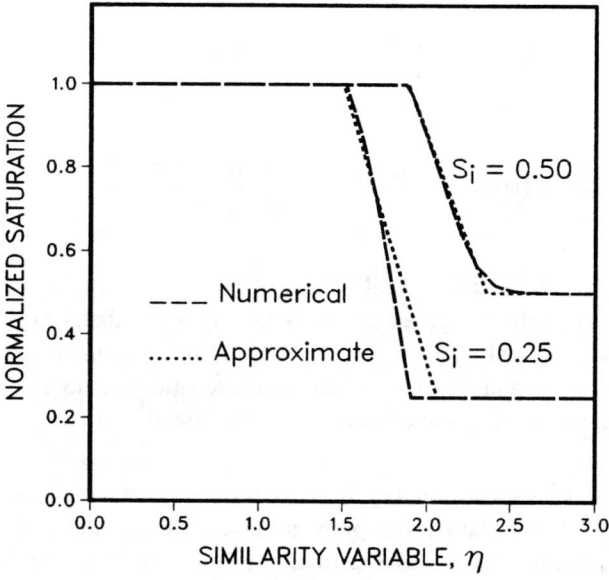

Figure 3. Saturation profiles for absorption into a Brooks-Corey medium, according to the approximate and numerical solutions.

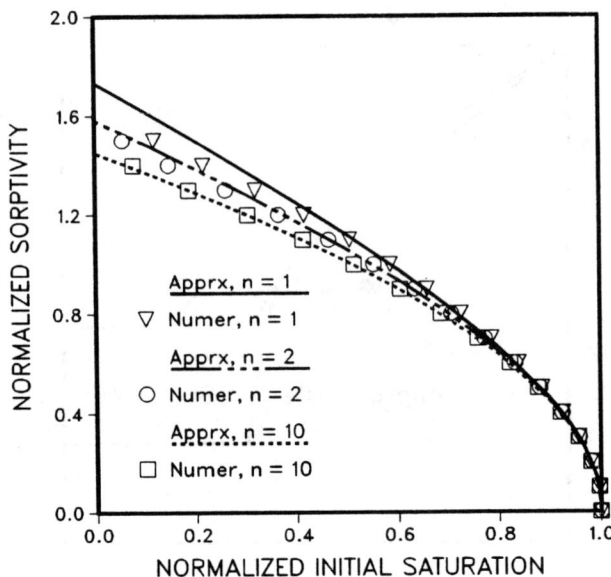

Figure 4. Sorptivity of a Brooks-Corey medium as a function of the initial saturation, for various values of the parameter n.

Solution for van Genuchten Media. Another commonly-used form for the characteristic curves of a porous rock or soil are those of van Genuchten [11] and Mualem [12]:

$$\hat{S}(\psi) = \frac{S(\psi) - S_r}{S_s - S_r} = [1 + (\alpha|\psi|)^n]^{-m} , \tag{22}$$

$$k_r(\psi) = \frac{\{1 - (\alpha|\psi|)^{n-1}[1 + (\alpha|\psi|)^n]^{-m}\}^2}{[1 + (\alpha|\psi|)^n]^{m/2}} . \tag{23}$$

The parameters S_s, S_r, and α have similar interpretations as they do in the Brooks-Corey expressions (6) and (7), while the parameter m is merely a shorthand expression for $1 - 1/n$. The parameter n should satisfy the inequality $n \geq 2$ [13]. The main qualitative difference between the Brooks-Corey and van Genuchten characteristic equations is that, in the latter case, the capillary pressure is a single-valued and continuous function of saturation in the region near $\hat{S} = 0$.

If the similarity transformation (8) is used on the governing equations, they again reduce to the two-point boundary-value problem given by Equations (9-11). In order to arrive at an appropriate form for the trial saturation profile, we first note that, for the flux of liquid into the medium to be finite, the slope $d\hat{\psi}/d\eta$ must be finite at $\eta = 0$. Since $\hat{\psi}(0) = 0$, this implies that $\hat{\psi} = -a\eta + \cdots$ for small values of η, where a is some constant. Substitution of this into Equation (22) shows that, to first order, $\hat{S} = 1 - ma^n\eta^n$. If we denote the length of the wetted zone by δ, a saturation profile of the form $\hat{S} = 1 - ma^n\eta^n$ will only satisfy the condition $\hat{S}(\delta) = \hat{S}_i$ if $ma^n = (1 - \hat{S}_i)/\delta^n$, so that

$$\hat{S}(\eta) = 1 - (1 - \hat{S}_i)\left[\frac{\eta}{\delta}\right]^n , \quad \text{for} \quad 0 < \eta < \delta ,$$

$$\hat{S}(\eta) = \hat{S}_i , \quad \text{for} \quad \delta < \eta < \infty . \tag{24}$$

Note that, as for the Brooks-Corey medium, we are using the simplest profile that is consistent with the boundary conditions and continuity requirements.

The parameter δ is found by integrating the governing equation (9) from $\eta = 0$ to $\eta = \infty$, with the saturation profile (24) substituted for $\hat{S}(\eta)$. The first term in Equation (9) integrates to

$$k_r(\hat{S}) \frac{d\hat{\psi}}{d\eta}\bigg|_0^\infty = a = \frac{(1 - \hat{S}_i)^{1/n}}{m^{1/n}\delta} . \tag{25}$$

The second term in Equation (9) integrates to

$$\int_0^\infty \frac{\eta}{2} \frac{d\hat{S}}{d\eta} d\eta = -\int_0^\delta \frac{\eta}{2} \frac{(1 - \hat{S}_i)n\eta^{n-1}}{\delta^n} d\eta = \frac{-(1 - \hat{S}_i)n}{2\delta^n} \int_0^\delta \eta^n d\eta$$

$$= \frac{-(1 - \hat{S}_i)n\delta}{2(n+1)} . \tag{26}$$

Combining Equations (25) and (26) leads to the following expression for δ:

$$\delta = \left[\frac{2(n+1)(1 - \hat{S}_i)^{-m}}{n(m^{1/n})}\right]^{1/2} . \tag{27}$$

Application of Darcy's law, as in Equation (19), along with the use of Equations (8) and (20), leads to the sorptivity in the form

$$\mathbf{S} = \left[\frac{2nk\phi(S_s - S_i)^{1+1/n}}{\alpha\mu[m(S_s - S_r)]^{1/n}}\right]^{1/2} . \tag{28}$$

As an example of the accuracy of this approximate solution, consider the problem of one-dimensional absorption from, say, a saturated fracture into the adjacent rock in the Topopah Spring unit at Yucca Mountain, Nevada, a potential site of an underground repository for high-level radioactive waste. This unit is a welded volcanic tuff whose hydraulic properties have been estimated [17] to be $\phi = 0.14$, $k = 3.9 \times 10^{-18} \, \mathrm{m}^2$, $S_s = 0.984$, $S_r = 0.318$,

$n = 3.04$, $m = 0.671$, and $\alpha = 1.147 \times 10^{-5}\,\text{Pa}^{-1}$. Consider the problem of absorption of water into a block of Topopah Spring welded tuff that is initially at a capillary pressure of -1 bar, which corresponds to an initial liquid saturation of 0.6765. If the temperature is taken to be 20 °C, then the viscosity of the water will be 0.001 Pa s (1 cp). The saturation profiles of the approximate and essentially exact (numerical) solutions after $1 \times 10^7\,\text{s}$ (116 days) of infiltration are shown in Fig. 5. Note that the approximate solution predicts the location of the wetting front extremely accurately, while the sorptivity, which is proportional to the area under the saturation curve, is overpredicted by a few percent. This is due to the fact that the while saturation follows a power-law profile near the boundary, this simple one-parameter expression (Equation (24)) does not represent the actual profile throughout the entire wetted zone with complete accuracy. However, the remarkable fact remains that a reasonably accurate approximate solution has been obtained, requiring neither extensive mathematical manipulations, nor any particular "physical insight" in order to arrive at the proper form for the saturation profile.

Conclusions. The integral method has been used to develop closed-form approximate solutions to the problem of water absorption into porous rock or soil. Solutions were developed for two widely-used forms of the capillary pressure and relative permeability equations, those of Brooks and Corey and van Genuchten-Mualem. The method requires only elementary integrations and differentiations, and leads to sorptivity predictions that are typically accurate to within better than 10%. In contrast to numerical solutions, the results of the integral method clearly display the manner in which the parameters of the problem affect the solution. Another point which was illustrated by these examples is that acceptable profiles can be found merely from consideration of boundary and continuity conditions. Other examples of the use of the integral method to find approximate solutions to porous media flow problems can be found in [18-20].

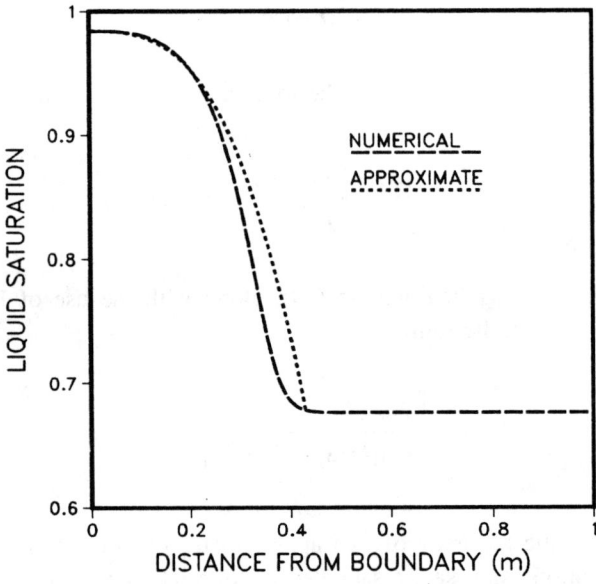

Figure 5. Saturation profile after 116 days of absorption into Topopah Spring welded tuff. The tuff is modeled as a van Genuchten medium; hydrological parameters are listed in text.

References

1. K. POHLHAUSEN, *Zur näherungsweisen Integration der Differentialgleichung der Grenzschicht*, Z. Ang. Math. Mech., 1 (1921), pp. 252-268. (In German).

2. H. BLASIUS, *Grenzschichten in Flüssigkeiten mit kleiner Reibung*, Z. Math. Phys., 56 (1908), pp. 1-37. (In German).

3. H. D. LANDAHL, *An approximation method for the solution of diffusion and related problems*, Bull. Math. Biophys., 15 (1953), pp. 49-61.

4. T. R. GOODMAN, *Application to integral methods to transient nonlinear heat transfer*, Adv. Heat Transf., 1 (1964), pp. 51-122.

5. J. R. PHILIP, *Numerical solution of equations of the diffusion type with diffusivity concentration-dependent*, Trans. Faraday Soc., 51 (1955), pp. 885-892.

6. D. HILLEL, *Fundamentals of Soil Physics*, San Diego, Academic Press, 1980.

7. M. KUTILEK, and J. VALENTOVA, *Sorptivity approximations*, Transp. Porous Media, 1 (1986), pp. 57-62.

8. J. R. PHILIP, *The theory of infiltration, 6, Effect of water depth over soil*, Soil Sci., 85 (1957), pp. 278-286.

9. J. BEAR, *Dynamics of Fluids in Porous Media*, Dover, New York, 1988.

10. R. H. BROOKS, and A. T. COREY, *Properties of porous media affecting fluid flow*, Proc. Amer. Soc. Civ. Eng. Irrig. Div., 92 (1966), pp. 61-87.

11. M. Th. VAN GENUCHTEN, *A closed-form equation for predicting the hydraulic conductivity of unsaturated soils*, Soil Sci. Soc. Amer. J., 44 (1980), pp. 892-898.

12. Y. MUALEM, *A new model for predicting the hydraulic conductivity of unsaturated porous media*, Water. Resour. Res., 12 (1976), pp. 513-522.

13. C. FUENTES, R. HAVERKAMP, J. Y. PARLANGE, W. BRUTSAERT, K. ZAYANI, and G. VACHAUD, *Constraints on parameters in three soil-water capillary retention equations*, Transp. Porous Media, 6 (1991), pp. 445-450.

14. R. R. BRUCE, and A. KLUTE, *The measurement of soil moisture diffusivity*, Soil Sci. Amer. Proc., 20 (1956), pp. 458-462.

15. W. BRUTSAERT, *The concise formulation of diffusive sorption of water in a dry soil*, Water Resour. Res., 12 (1976), pp. 1118-1124.

16. W. H. PRESS, B. P. FLANNERY, S. A. TEUKOLSKY, and W. T. VETTERLING, *Numerical Recipes*, New York, Cambridge University Press, 1986.

17. J. RULON, G. S. BODVARSSON, and P. MONTAZER, *Preliminary numerical simulations of groundwater flow in the unsaturated zone, Yucca Mountain, Nevada*, Report LBL-20553, Lawrence Berkeley Laboratory, Berkeley, Calif., 1986.

18. S. N. PRASAD and M. J. M. ROMKENS, *An approximate integral solution of vertical infiltration under changing boundary conditions*, Water Resour. Res., 18 (1982), pp. 1022-1028.

19. R. W. ZIMMERMAN, and G. S. BODVARSSON, *Integral method solution for diffusion into a spherical block*, J. Hydrol., 111 (1989), pp. 213-224.

20. R. W. ZIMMERMAN, G. S. BODVARSSON, and E. M. KWICKLIS, *Absorption of water into porous blocks of various shapes and sizes*, Water Resour. Res., 26 (1990), pp. 2797-2806.